AF184128

SEMA SIMAI Springer Series

Volume 40

Editors-in-Chief

Carlos Castro, Departamento de Matemática e Informática, ETSI Caminos, Canales y Puertos, Universidad Politécnica de Madrid, Madrid, Spain

Luca Formaggia, MOX–Department of Mathematics, Politecnico di Milano, Milano, Italy

Series Editors

Maria Groppi, Dipartimento di Scienze Matematiche, Fisiche e Informatiche, Università di Parma, Parma, Italy

Mats G. Larson, Department of Mathematics, Umeå University, Umeå, Sweden

Maria Lopez Fernandez, Department of Mathematical Analysis, Statistics and O.R., and Applied Mathematics, Faculty of Sciences, University of Malaga, Malaga, Spain

Tomás Morales de Luna, Departamento de Análisis Matemático, Estad. e I.O., y Matemática Aplicada, Facultad de Ciencias, Universidad de Málaga, Málaga, Spain

Lorenzo Pareschi, Dipartimento di Matematica e Informatica, Università degli Studi di Ferrara, Ferrara, Italy

Elena Vázquez-Cendón, Departamento de Matemática Aplicada, Universidade de Santiago de Compostela, A Coruña, Spain

Paolo Zunino, Dipartimento di Matematica, Politecnico di Milano, Milano, Italy

As of 2013, the SIMAI Springer Series opens to SEMA in order to publish a joint series aiming to publish advanced textbooks, research-level monographs and collected works that focus on applications of mathematics to social and industrial problems, including biology, medicine, engineering, environment and finance. Mathematical and numerical modeling is playing a crucial role in the solution of the complex and interrelated problems faced nowadays not only by researchers operating in the field of basic sciences, but also in more directly applied and industrial sectors. This series is meant to host selected contributions focusing on the relevance of mathematics in real life applications and to provide useful reference material to students, academic and industrial researchers at an international level. Interdisciplinary contributions, showing a fruitful collaboration of mathematicians with researchers of other fields to address complex applications, are welcomed in this series.

THE SERIES IS INDEXED IN SCOPUS

Michael Stich · Jorge Carballido-Landeira
Editors

Nonlinear Dynamics for Biological Systems

Springer

Editors
Michael Stich
Departamento de Matemática Aplicada,
Ciencia e Ingeniería de los Materiales y
Tecnología Electrónica
Universidad Rey Juan Carlos
Móstoles, Madrid, Spain

Jorge Carballido-Landeira
Departamento de Física
Universidad de Oviedo
Oviedo, Asturias, Spain

ISSN 2199-3041 ISSN 2199-305X (electronic)
SEMA SIMAI Springer Series
ISBN 978-3-031-99043-4 ISBN 978-3-031-99044-1 (eBook)
https://doi.org/10.1007/978-3-031-99044-1

© The Editor(s) (if applicable) and The Author(s), under exclusive license to Springer Nature Switzerland AG 2025

This work is subject to copyright. All rights are solely and exclusively licensed by the Publisher, whether the whole or part of the material is concerned, specifically the rights of translation, reprinting, reuse of illustrations, recitation, broadcasting, reproduction on microfilms or in any other physical way, and transmission or information storage and retrieval, electronic adaptation, computer software, or by similar or dissimilar methodology now known or hereafter developed.
The use of general descriptive names, registered names, trademarks, service marks, etc. in this publication does not imply, even in the absence of a specific statement, that such names are exempt from the relevant protective laws and regulations and therefore free for general use.
The publisher, the authors and the editors are safe to assume that the advice and information in this book are believed to be true and accurate at the date of publication. Neither the publisher nor the authors or the editors give a warranty, expressed or implied, with respect to the material contained herein or for any errors or omissions that may have been made. The publisher remains neutral with regard to jurisdictional claims in published maps and institutional affiliations.

This Springer imprint is published by the registered company Springer Nature Switzerland AG
The registered company address is: Gewerbestrasse 11, 6330 Cham, Switzerland

If disposing of this product, please recycle the paper.

Preface

Mathematics is fundamental for describing the structure and dynamics of Nature in a formal language. At all times, mathematics was crucial for the cultural and technological development of human societies: While still not called as such at the time, astronomy was probably the first science to use mathematical methods systematically. Observations of celestial bodies like the Sun or the Moon provided the basis for analysis and subsequent predictions for the change of the seasons, the migration of animals, the availability of water resources and, eventually, a measure of time. Beyond the practical use of ensuring survival, the possibility of making at least some predictions about the future was accompanied by a spiritual dimension and provided power to those able to use symbolic methods and reasoning. Without even trying to attempt a detailed historical account, we still note that over time and across the globe, natural philosophy developed against the background of different cultural, religious, and philosophical ideas. Slowly, the natural sciences now known as physics and chemistry shaped their empirical and methodological basis, left the realm of a mere qualitative description, and became areas of abstract reasoning, successful mathematical analysis, modelling, and prediction.

Despite these accomplishments, other sciences have resisted longer to take the path towards quantification and mathematical analysis beyond descriptive statistics. Biology, the science of the living being, is one of them. Already one of the simplest biological entities, the single cell, is fundamentally different from inanimate matter and difficult to model mathematically if all aspects of life should be represented and incorporated. The reductionist principle, so fruitful in the physical sciences, is constantly challenged by the high level of complexity we observe in biological systems: for example, no equivalent to the simple linear oscillator, so crucial for modelling a wide range of mechanical, quantum-mechanical or electrodynamical processes, seems to exist! Obviously, the breakdown of the reductionist principle is not exclusive for biological systems, it is also true for other *complex systems*.

The complexity of a biological system is the result of a highly diverse molecular composition, intricate biochemical reaction networks, and structures being organized from the molecular length scale up to the level of organisms. Furthermore, feedback loops and non-trivial temporal dynamics operate on these different length scales, enable interaction with other living and non-living matter, create emergent properties, and are subject to aging on an individual level and to evolutionary processes on the population level due to the existence of genes and the transmission of information from generation to generation.

One of the most important and successful mathematical methodologies in the modelling of such dynamical and evolutionary systems is *nonlinear dynamics*. As for other areas of applied mathematics, nonlinear dynamics does not represent a closed set of methods or equations but is constantly developing and redefining itself due to the diverse systems it is applied to. At the heart of nonlinear dynamics are mappings or sets of nonlinear difference or differential equations, and mathematical questions typically address the existence of solutions, their unicity, and their stability with respect to initial conditions or parameter changes. Related research fields are dynamical systems theory, network and graph theory, chaos theory, complexity science, control theory, systems biology and even bioinformatics, to name a few. All these methods can be applied successfully to deepen our understanding of biological systems! And from there is—depending on the specific context—only one step to the application of such methods in medicine.

From 6th to 8th of June 2022, the 3rd workshop on Nonlinear Dynamics in Biological Systems took place in Madrid. It united researchers working on the theoretical and data-analytical aspects of biological systems with biologists and experimental biophysicists, with the central aims of learning from each other and of extending our knowledge of a range of biological systems, in particular, with methods related to nonlinear dynamics. The sessions of this workshop are reflected by the selection of sections and chapters for this book, which were contributed by participants of the workshop.

In the first part of this book, three chapters review recent results in the field of neurodynamics and modelling of neuronal systems. The contributions unravel the dynamics of sensorimotor integration (Chapter "The Dynamics of Sensorimotor Integration"), study the intrinsic dynamics of neural networks (Chapter "Unveiling the Intrinsic Dynamics of Biological and Artificial Neural Networks: From Criticality to Optimal Representations") and investigate emergent complex dynamics in neuronal cultures and its application to medicine (Chapter "Emergent Complex Dynamics in Neuronal Cultures and Its Relation to Neuroengineering and Medicine"). In the second part, two chapters are dedicated to the dynamics of cells: the first contribution is devoted to branching processes to study stem cell differentiation dynamics (Chapter "Modeling and Analyzing Biological Systems Using Branching Processes") and the second one to quantitative model evaluation in stochastic reaction-diffusion systems in biophysics (Chapter "Stochastic Reaction-Diffusion Systems in Biophysics: Towards a Toolbox for Quantitative Model Evaluation"). And finally, in the third part we present chapters that focus on network aspects and the interactional nature of biological systems, namely network and beyond-network approaches in evolution (Chapter "Networks: The Visual Language of Complexity"), a lung model for tuberculosis (Chapter "Modelling the Formation of the Granuloma Inside Secondary Lobule of the Lungs") and sustainable biodegradation process of new organic semiconductors (Chapter "Mathematical Model for Sustainable Biodegradation Process"). We hope the readers enjoy the travel through these exciting research fields!

The workshop and this book would not have been possible without the financial support of the following institutions: first, and foremost, the Universidad Rey Juan Carlos as the host of the workshop, but also the Berlin Center for Studies of Complex Chemical Systems (BCSCCS), the German Sonderforschungsbereich Data Assimilation (SFB 1294), the Spanish Ministerio de Ciencia e Innovación via project PID2020-113737GB-I00 and the journal Chaos, Solitons and Fractals through providing a travel grant. We also acknowledge support from the Spanish Network "Nonlinear Dynamics in Biophysical Systems" (AEI).

Móstoles, Spain Michael Stich
Oviedo, Spain Jorge Carballido-Landeira

Contents

About the Editors

Michael Stich is currently an Associate Professor in Mathematics at the University King Juan Carlos in Madrid, Spain. His research interests include mathematical modelization of dynamical biological and chemical systems, in particular the modelling of RNA and molecular evolution, antimicrobial resistance, and the origin of life. He is member of the Spanish network "Nonlinear Dynamics in Biophysical Systems".

Jorge Carballido-Landeira is currently an Associate Professor in the Department of Applied Physics at the University of Oviedo, Spain. His research focuses on a multidisciplinary field that includes nonlinear dynamics, self-organization processes, colloidal and polymer science, environmental constraints and fluid mechanics for gaining an understanding of complex biological systems.

Neuronal Dynamics

The Dynamics of Sensorimotor Integration

Agustín Carpio Andrada and Gabriel B. Mindlin

Abstract Sensorimotor integration refers to how the sensory feedback originating from motor actions influences the generation of motor patterns. From a dynamic perspective, this can be understood as the impact of a term in the equations governing motor behavior, which is dependent on the variables of the problem in previous instances. An important motor pattern is the generation of periodic behavior, and in this work, we will discuss the impact of delayed feedback in a system capable of displaying these patterns. This question has been addressed when the appearance of periodic motor patterns is due to a Hopf bifurcation. Here we review those results, and then move to explore the rich emergent dynamics arising in delayed systems near a Saddle-Node In Limit Cycle (SNILC) bifurcation. Our results reveal a complex subharmonic structure consistent with known activity patterns in multiple fields. We also explore potential applications of this dynamic phenomenon.

1 Introduction

Over the past few decades, extensive research has been conducted to comprehend the neural control of movement. By employing straightforward computational models to depict the actions of a single neuron or the collective actions of a neural population, it became feasible to construct basic networks capable of generating a diverse range of spatio-temporal patterns. These can be interpreted as the physiological instructions controlling the biomechanics of some peripheral system.

Certain networks, characterized as dynamical systems, exhibit the ability to yield diverse solutions contingent on specific global parameters. This capability has proven instrumental in shedding light on various phenomena, notably contributing to our

A. Carpio Andrada · G. B. Mindlin (✉)
Departamento de Física, Facultad de Ciencias Exactas y Naturales, Ciudad Universitaria, Universidad de Buenos Aires, Buenos Aires, Argentina
e-mail: gabo@df.uba.ar

Instituto de Física Interdisciplinaria y Aplicada (INFINA), Ciudad Universitaria, CONICET - Universidad de Buenos Aires, Buenos Aires, Argentina

© The Author(s), under exclusive license to Springer Nature Switzerland AG 2025
M. Stich and J. Carballido-Landeira (eds.), *Nonlinear Dynamics for Biological Systems*, SEMA SIMAI Springer Series 40,
https://doi.org/10.1007/978-3-031-99044-1_1

understanding of distinct gaits observed in locomotion. In the 1970s, physiologists drew inspiration from engineering control systems to gain insight into processes involving feedback, thereby enhancing the comprehensive description of behavior.

The behavior of an animal, for instance when it is executing a motor plan, emerges from the intricate interplay among its nervous system, its motor morphological structures, and the sensory information gathered from its environment. This interaction operates bidirectionally, as the environment is also influenced by the animal's behavior. The concept of "sensory feedback" encompasses the mechanism through which animals perceive and assess the consequences of their actions on the environment. In parallel, "sensorimotor integration" involves the process by which motor output, or behavior in general, materializes through the interplay between two information channels: sensory input processing and motor plan pathways [20].

The research program is simple to articulate, yet its implementation is not without challenges, primarily due to the intricacies involved in deciphering the coding of sensory information. Within the framework of functionalism in neuroscience, the nervous system processes sensory input based on its intended purpose and destination. When the goal is perception, the processing may entail feature extraction and the amalgamation of inputs from various sensory sources. Conversely, if the sensory input is employed to guide the generation of motor commands, it must be interpreted in relation to the present state of the motor system and how it may change the course of execution of motor plans.

Repetitive sequencing of actions is a prevalent phenomenon in behavioral sequences, particularly observed in animal vocalizations. Notably, many species of birds construct their songs through the iterative repetition of specific syllables, followed by a sequence featuring a distinct syllable. Elementary models elucidating birdsong production reveal that these reiterated syllables can be produced through periodic physiological instructions governing respiration and the configuration of the vocal organ. The generation of different syllables can be achieved by modifying the relative phases of these instructions or other features such as their periods.

Extensive research has been dedicated to the avian song system, with a consensus emerging that a specific brain area within the telencephalon, known as HVC (proper name, formerly stood for high vocal center), plays a pivotal role in temporal pattern generation. Moreover, it is established that the auditory pathway projects into this brain region. Consequently, it is reasonable to inquire whether auditory feedback plays a crucial role in sustaining syllable repetitions in certain bird species. Corroborating this hypothesis, experiments with Bengalese finches indicate that deafening leads to a reduction in syllable repetitions. For the case of birdsong, the study of a computational model showed that syllable repeats are initially sustained by auditory feedback [20]. However, periodic patterns of activity are pervasive in nature and manifest in a variety of animal behaviors [19, 21]. The hypothesis posits that, in certain contexts, rhythmic activity emerges from the collective behavior of neural oscillators, which generate periodic signals driving the animal's motor systems [10].

A neural oscillator, a simple neural architecture capable of instigating periodic activity within the brain, comprises a pair of interconnected excitatory and inhibitory neurons [6]. When a local population of neurons exhibits comparable dynamics and functionality, extending the neural oscillator model from a pair of neurons to a pair of populations using mean field theory becomes advantageous [8, 15]. The comprehension of these oscillators' dynamics and their interaction with other components of the nervous system, such as those conveying sensory feedback, is crucial for unraveling the mechanisms underpinning the generation of periodic behavior. Mathematically, this endeavor necessitates a thorough review of the impact of incorporating a delayed function of the variables used to describe the problem in the model elucidating the origin of the oscillations.

This issue has been previously investigated in [14], where the impact of delayed feedback on a basic oscillator is explored. This seminal work modeled the oscillator as the normal form of a Hopf bifurcation, which is a simple representation of how oscillations arise in nonlinear systems. In a Hopf bifurcation, oscillations emerge with zero amplitude and a distinct frequency. Building upon this foundation, our study extends the previous research by examining the effects of delayed feedback in proximity to global bifurcations. Furthermore, we investigate this problem within the framework of an interpretable model for neural oscillators.

In computational neuroscience, the Wilson-Cowan model [18] delineates the dynamics governing interactions among populations of elementary excitatory and inhibitory model neurons. Originating from the collaborative work of Hugh R. Wilson and Jack D. Cowan, this model has found widespread application in neuronal population modeling, with various adaptations gaining prominence. Notably, the model holds historical significance for its utilization of phase plane methods and numerical solutions to elucidate the responses of neuronal populations to stimuli.

A distinctive feature of the Wilson-Cowan model is its depiction of the interplay between a population that stimulates the neurons it connects to and a population that inhibits its efferents (the neurons it connects to). As a consequence of this interaction, a common dynamical outcome is the emergence of oscillations. Oscillations within a Wilson-Cowan model can emerge not only through local bifurcations like Hopf bifurcations but also through global bifurcations, exemplified by the Saddle-Node In Limit Cycle bifurcation (SNILC). In this type of bifurcation, a local Saddle-Node bifurcation takes place, but since the unstable manifold of the saddle is part of the stable manifold of the attractor, the disappearance of the fixed points gives rise to a periodic solution with infinite period at the bifurcation (for this reason, SNIPER is an alternative name for this bifurcation; it stands for Saddle Node in Infinite Period). These bifurcation types are closely related to neuronal excitability classes, such as Type I and Type II, as described in [11]. Notice that the values of the vector field in the region of the phase space where the two fixed points collapsed will be close to zero. Therefore, the periodic trajectory will be slow (critical slowing down phenomena) in that region of the phase space. Consequently, the oscillations exhibit a composite structure, comprising both fast and slow components. This distinctive feature sets them apart from simple harmonic oscillations. Another consequence is that the periodic solutions being born in these bifurcations contain a rich spectral content.

The richness of the dynamics in the Wilson-Cowan oscillator suggests that to comprehend the integration of a sensorimotor effect, translated mathematically through the inclusion of a delay term, one must delve beyond the impact of such terms on a simple Hopf oscillator. For this reason, in this chapter, we will scrutinize the bifurcations inherent in the dynamics of a Wilson-Cowan oscillator and explore the repercussions of introducing delayed terms to the model when it is close to a SNILC bifurcation. By incorporating these delays, our goal is to delve into the potential dynamics that emerge when the onset of periodic motor control integrates with sensory feedback. The inclusion of delay in this context signifies the cumulative time required for motor behavior to undergo sensory integration and subsequently feed back into the motor program.

This chapter is structured as follows: Sect. 2 revisits a study conducted by Ramana Reddy et al. [14] that investigates the effects of time-delayed linear and nonlinear feedback on the dynamics of a single Hopf bifurcation oscillator. In Sect. 3, we present a comprehensive analysis of the proposed model, including a phase plane analysis that examines various dynamical regimes. Furthermore, we provide a complete bifurcation map of the system when the delayed term is omitted. Section 4 explores the influence of delayed feedback on the model. We construct a map that illustrates the range of periodic solutions in a delay-parameter space (a $K \times \tau$ space where K represents the amplitude of a delayed feedback and τ its delay-time lag). Our findings reveal increased complexity due to the introduction of a delay-term exhibiting period-doubling bifurcations, hysteresis, and phase-locking to a spectrum of subharmonic solutions. Finally, Sect. 5 summarizes the key findings and implications of our study. Additionally, we discuss future research directions, including a biological perspective on oscillatory neural activity and periodic patterns observed in animal behavior.

2 The Effect of Feedback on Systems Close to Hopf and Saddle-Node-in-Limit-Cycle Bifurcations

Normal forms are the simplest equations consistent with a linear singularity, where simple refers to the minimal number of monomial terms in the vector field up to a given order. For the case of a Hopf bifurcation, the qualitative change in the flow being the birth of a limit cycle of zero amplitude and finite frequency, the normal form for a complex variable depending on time $Z(t)$ reads as follows:

$$\dot{Z}(t) = (a + i\omega - |Z(t)|^2)Z(t), \tag{1}$$

where a is a real constant and ω denotes the frequency at which oscillations arise in the absence of feedback. For its simplicity, it serves as a widely utilized dynamical system for representing periodic behavior. In [14], the authors introduce a feedback term to the system as follows:

$$\dot{Z}(t) = (a + i\omega - |Z(t)|^2)Z(t) - KZ(t - \tau), \tag{2}$$

where K represents the amplification factor of the delayed term, and τ corresponds to the time delay. The delayed feedback term, $KZ(t-\tau)$, accounts for the contribution of the system's activity at time $t-\tau$ to its behavior at time t, weighted by the factor K. When $K=0$, the system exhibits a stable limit cycle with an amplitude of $\sqrt{a}$ and an angular frequency of ω.

By expressing the system in polar coordinates and setting $a=1$, we can rewrite the equations as:

$$\dot{r}(t) = [1 - r^2(t)]r(t) - Kr(t-\tau)\cos[\theta(t-\tau) - \theta(t)] \tag{3}$$

$$\dot{\theta}(t) = \omega - K\frac{r(t-\tau)}{r(t)}\sin[\theta(t-\tau) - \theta(t)]. \tag{4}$$

Considering solutions of the form $Z(t) = Re^{i\Omega t}$ for $\tau > 0$, we anticipate solutions in polar form as $r(t) = R$ and $\theta(t) = \Omega t$, where R and Ω are real constants. Thus, we can deduce the mathematical conditions for the amplitude and frequency of the oscillator in the periodic solution:

$$R = \sqrt{1 - K\cos(\Omega\tau)} \tag{5}$$

$$\Omega = \omega + K\sin(\Omega\tau) \tag{6}$$

Graphically, Eq. (6) reveals that the frequencies of the solutions correspond to the intersection points of the identity function $f(\Omega) = \Omega$ and $g(\Omega) = \omega + K\sin(\Omega\tau)$. Under specific conditions of ω and τ, multiple intersections exist, indicating the presence of multiple periodic solutions. An example demonstrating this phenomenon is illustrated in Fig. 1.

Indeed, the inclusion of feedback imposes a significant constraint on the frequencies at which periodic oscillations can exist. This is in stark contrast to the situation without feedback, where a limit cycle can exist for any value of Ω. The introduction of delayed feedback introduces a condition that restricts the possible frequencies at which sustained oscillations can occur. This constraint arises from the interplay between the intrinsic frequency of the system, represented by ω, and the influence of the delayed feedback term. As a result, the system's dynamics exhibits a more intricate behavior, characterized by a limited range of frequencies that support sustained oscillations.

In situations where multiple stable solutions coexist for a given parameter set, the system's evolution towards a specific stable solution depends on the initial condition. These initial condition can be conceptualized as a vector with infinitely many complex elements, representing the system's states for every t between $-\tau$ and 0 (if we start integrating from $t=0$).

Numerical simulations have confirmed the presence of hysteresis phenomena (dependence of the state of a system on its history) associated with the existence of multiple stable periodic solutions. Depending on the initial condition, the system may converge towards one particular stable solution or another, leading to distinct

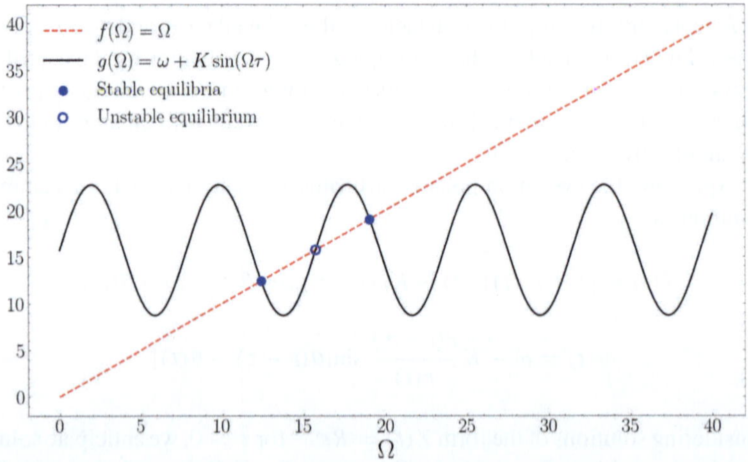

Fig. 1 Periods of rhythmic solutions in Eq. (1) are given by the intersection of $f(\Omega)$ and $g(\Omega)$. In this example, two stable and one unstable solutions exist with parameters $\omega = 15.7$, $K = 7$, $\tau = 0.8$

dynamical behaviors and trajectories. This behavior highlights the sensitivity of the system to its initial state and the rich complexity arising from the coexistence of multiple stable solutions.

The study of under which conditions a system's vector field undergoes a topological change can be addressed by bifurcation theory. For a two-dimensional dynamical system described by its associated Jacobian matrix (the matrix of all the system's first-order partial derivatives), specific conditions determine the occurrence of the Saddle-Node and Hopf bifurcations. The Saddle-Node bifurcation condition is given by $D = 0$, where D represents the determinant of the Jacobian matrix. When $D = 0$, a Saddle-Node bifurcation occurs, resulting in the creation or destruction of a pair of equilibrium points.

On the other hand, the Hopf bifurcation condition is satisfied when $T = 0$ and $D > 0$, leading to the birth or annihilation of a limit cycle, where T stands for the trace of the Jacobian matrix. These conditions provide insights into the behavior and qualitative changes that occur in the system as its parameters are varied, helping to understand the occurrence of bifurcations and the emergence of different dynamical regimes.

Unlike the Hopf bifurcation, which typically gives rise to periodic solutions with a well defined finite frequency, the SNILC bifurcation exhibits a distinct behavior characterized by the emergence of periodic solutions which are born with infinite period. This results in a diverse set of periodic behaviors that span a broad range of time scales [4] for parameters in the vicinity of the bifurcation. This wide range of periods adds richness and complexity to the system dynamics, offering a broader repertoire of possible behaviors compared to other bifurcations.

Notice that the SNILC bifurcation is global in nature. Hence, we will explore the effect of feedback on a simple and pertinent system of equations that, with minimal complexity, is able to describe neural oscillations born in SNILC bifurcations.

Now that we know some of the features that usually emerge in delayed feedback systems, let's take a step back to gently introduce the Wilson-Cowan neural oscillator model. In Sect. 3 we present the system and describe it in terms of bifurcation theory for the ordinary form (i.e. with the delay term multiplied by zero). Then in Sect. 4, we explore in depth the effects of a delayed copy of the system's activity feeding back into itself while varying the delay related parameters. The use of this set of equations allows a smooth export to the realm of neuroscience, where sensorimotor integration of major importance.

3 The Dynamics of Neural Oscillators

We start reviewing the solutions of a Wilson-Cowan oscillator, which is a system designed to describe the dynamics of two interacting populations of neurons: excitatory and inhibitory. Our goal is to modify the model in order to incorporate the effect of feedback, which can be interpreted as the sensory processing of changes resulting from the system's own activity (i.e., sensory feedback). To achieve this, we introduce a delayed copy of one of the variables as an input to the system:

$$\begin{cases} \dot{x} &= \mu(-x + S(\rho_x + ax - by)) \\ \dot{y} &= \mu(-y + S(\rho_y + cx - dy + Kx(t-\tau))) \end{cases}. \tag{7}$$

The state variables $x = x(t)$ and $y = y(t)$ represent the excitatory and inhibitory neural activities, respectively. The parameter μ governs the intrinsic time scale of the population's average firing rate. The sigmoid function $S(x) = \frac{1}{1+e^{-x}}$ captures the processing of inputs from both populations, weighted by synaptic strength coefficients a, b, c, and d, as well as external stimuli ρ_x and ρ_y. The time-delayed copy of the excitatory activity $x(t-\tau)$, serves as an input to the inhibitory population at time t. This accounts for the feedback effect on the neural dynamics in our model. This delayed input is multiplied by a factor K, representing an amplification factor. Note that the linear copy $x(t-\tau)$ is the simplest way to introduce delayed feedback into the system. Figure 2 provides a graphical representation of the described model.

Consider $K = 0$ (the ordinary non-delayed version of the system). In this case, Eq. (7) simply reduces to:

$$\begin{cases} \dot{x} &= \mu(-x + S(\rho_x + ax - by)) \\ \dot{y} &= \mu(-y + S(\rho_y + cx - dy)) \end{cases}. \tag{8}$$

This dynamical system serves as a model for the emergence of periodic patterns in the absence of sensory feedback. The condition for equilibrium is given by both

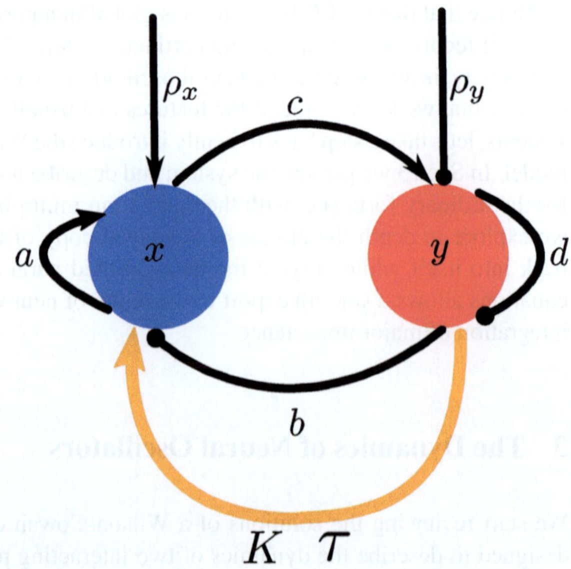

Fig. 2 Illustration of the delayed Wilson-Cowan model depicting excitatory (blue circle) and inhibitory (red circle) populations of neurons, and their activities denoted by x and y respectively. The synaptic parameters of the model, denoted by a, b, c, and d, as well as the input parameters ρ_x and ρ_y, are shown. Additionally, the delay-related parameters K and τ are indicated

derivatives being equal to zero. So, a fixed point defined as $P^* = (x^*, y^*)$ must satisfy the condition:

$$\begin{cases} x^* &= S(\rho_x + ax - by) \\ y^* &= S(\rho_y + cx - dy) \end{cases}. \tag{9}$$

Therefore, the Jacobian matrix corresponding to Eq. (8), denoted by J, which evaluated at the fixed points reads as:

$$J = \begin{bmatrix} -1 + aS'(\rho_x + ax - by) & -bS'(\rho_x + ax - by) \\ cS'(\rho_y + cx - dy) & -1 - dS'(\rho_y + cx - dy) \end{bmatrix}. \tag{10}$$

J can be expressed in a more convenient form as shown, addressing the fact that the derivative of the sigmoid function, S', is equal to $S(1-S)$ (full details on [6]):

$$J = \begin{bmatrix} -1 + ax(1-x) & -bx(1-x) \\ cy(1-y) & -1 - dy(1-y) \end{bmatrix}. \tag{11}$$

The Saddle-Node bifurcation occurs when the determinant of the evaluated Jacobian matrix becomes zero. However, the impact of this bifurcation on the system depends on factors beyond the mere change in sign of the determinant.

The occurrence of a Saddle-Node bifurcation does not necessarily imply the creation or destruction of a limit cycle. Determining whether a limit cycle is formed or destroyed as a result of the bifurcation depends on a comprehensive analysis of the system's global behavior, taking into account other components and factors. This particular case of Saddle-Node bifurcation is referred to as Saddle-Node in Limit Cycle bifurcation (SNILC).

In this section we describe the bifurcation diagram (a plot where multiple bifurcation curves are depicted in a parameter space, in this case the $\rho_x \times \rho_y$ space) of the model for the case of $K = 0$ and specific parameter values ($a = b = c = 10$, $d = -10$). These particular parameter values are selected to ensure that the boundaries of the oscillatory regime in the parameter space correspond to SNILC bifurcation curves and none of them to Hopf bifurcations.

Phase portraits provide a concise graphical representation of the qualitative behavior of the solutions of a system under specific parameter values [16]. When the system undergoes a bifurcation as a result of a change of parameters, the corresponding phase portraits exhibit distinct qualitative characteristics.

In our analysis, we will focus on the following features depicted in the phase portraits:

- The positions of fixed points
- The presence of closed orbits or limit cycles
- Trajectories of the system, which provide insights into the stability of fixed points and the direction of flow within the system.

As shown in Fig. 3, the central region of the parameter space displays distinctive periodic behavior characterized by limit cycle dynamics. This region is enclosed by SNILC bifurcation curves. In the proximity of the bifurcation, the period of a solution gradually increases as we approach the bifurcation point. Upon crossing the SNILC bifurcation, the system transitions into a quiescent regime, where sustained oscillations cease to exist.

Phase portraits were obtained by numerically integrating the system and analytically finding the fixed points. A representative phase portrait that captures the qualitative dynamics is drawn for each of the relevant parameter space regions delimited by bifurcation curves given by the parameters of that region. To provide a comprehensive understanding of the various dynamic regimes, we conducted an exploration along both local and global bifurcation curves.

In Fig. 3 we show both the local bifurcation curves and the global ones (including the homoclinic bifurcation curve depicted in blue). It is worth mentioning that some of the black curves depicting Saddle-Node bifurcations also correspond to global bifurcations, since some of them are SNILC.

So far, we have focused on the qualitative differences observed in the system's dynamics, evaluated in different regions of the $\rho_x \times \rho_y$ parameter space. However, it is important to note that in the vicinity of a bifurcation leading to qualitative changes, there exists a spectrum of quantitatively distinct solutions. Specifically, when the system is close to a SNILC bifurcation and a simple limit cycle is present, the periods of the solutions span a wide range even with small variations in the parameters. This is illustrated in Fig. 4, where we observe the variation in periods for different parameter values.

Having completed a comprehensive qualitative analysis of the dynamic regimes in the system, we are now poised to delve into the investigation of the effects of

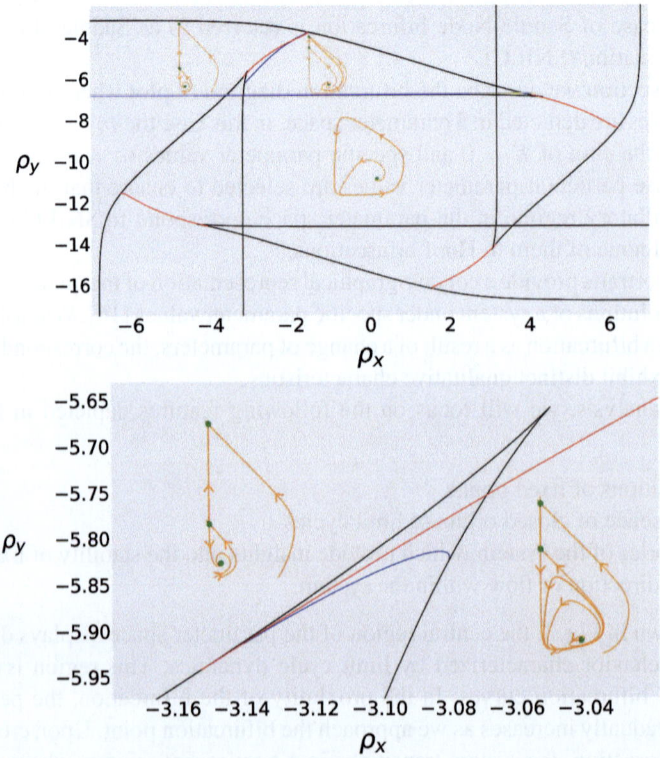

Fig. 3 Wilson-Cowan bifurcation map. Lower panel is a zoom of the top panel. The Saddle-Node bifurcation curves are represented in black, the Hopf bifurcation curves in red and the Homoclinic bifurcation curves in blue. Phase portraits (in orange) corresponding to the different delimited regions are represented inside the regions. Qualitatively equal phase portraits (as a consequence of symmetry) and those with 5 fixed points were omitted. It should be noted that, for simplicity, neither the node-spiral transitions nor the double limit cycle bifurcation has been represented in the map. Likewise, the parts of the Saddle-Node bifurcation that are also SNILC have not been distinguished, nor have the subcritical and supercritical parts of the Hopf bifurcation

delayed feedback. By incorporating this additional factor into the problem, we aim to gain insights into how it influences the system's behavior. This next section will explore the intriguing interplay between delayed feedback and the existing dynamic regimes, shedding light on the complex dynamics that emerge from this interaction.

4 The Effect of Feedback on Neural Oscillators

In this section, we introduce feedback into the neural oscillator model, following a similar approach as in previous studies [12–14]. By incorporating a non-zero value for the feedback parameter K in Eq. (7), the system becomes sensitive to its own

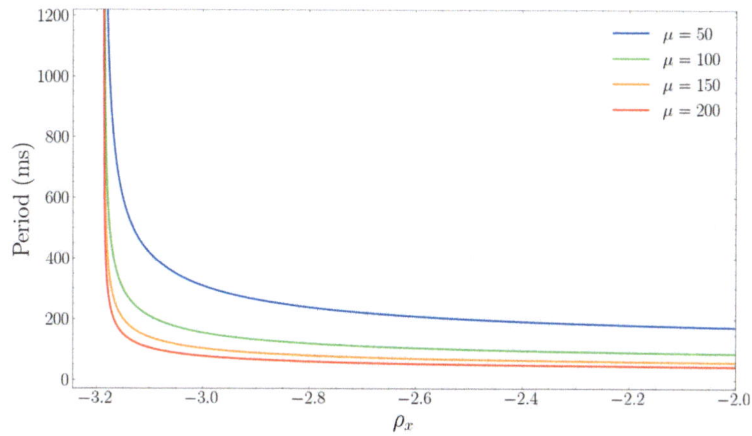

Fig. 4 Period against ρ_x for multiple values of μ in Eq. (7) with $a = b = c = 10, d = -10, \rho_y = -9$ and $K = 0$. Note that solutions' periods tend to infinity when ρ_x tends to the critical value for the SNILC bifurcation $\rho_x' \approx -3.2$

past states. To capture this effect computationally, we employ a queue-like structure to store the system states from $t - \tau$ to t, where τ represents the time delay. This structure is updated at each step of the numerical integration, which is performed using a fourth-order delay-adapted Runge-Kutta method. Additionally, we include an option to consider the state vector from the immediately previous computation in consecutive simulations, providing continuity in the analysis.

For the subsequent analysis, we will fix the non-delay parameters of Eq. (7) as described in the caption of Fig. 5, which guarantees the proximity to a SNILC. We then proceeded to explore the dynamics as K and τ are varied.

In the absence of feedback ($K = 0$), the system exhibits periodic solutions with a period duration of 65 ms. To explore the effects of feedback, we employed the modified fourth-order Runge-Kutta method discussed earlier to numerically compute the trajectories of the system. By integrating the equations over a wide range of K and τ values, we obtained the corresponding trajectories and investigated their characteristics. To comprehensively analyze the impact of feedback, we explored a subset of the $K \times \tau$ parameter space, where K ranges from 3 to 10 and τ ranges from 40 to 70 ms. By doing so, we aim to uncover the emerging complexity in the delayed version of the model.

The incorporation of a delay allows us to retain and analyze the previous states of the system in a "queue" structure. With this capability, we can traverse the $K \times \tau$ space in all four directions (ascending and descending values of K and τ, while keeping the other parameter constant) and investigate how these variations lead to the generation of qualitatively and/or quantitatively different periodic solutions. By examining the trajectories obtained for each K and τ combination, we can gain insights into the nature and diversity of the system's dynamics.

It is worth noting that the analysis of the solutions is performed disregarding any transients that may appear when increasing or decreasing a parameter between consecutive integrations. Since the length of the initial condition vector grows for consecutive integrations when τ is increasing, the stored states of the system must be long enough to avoid running short of stored values for $x(t-\tau)$ and $y(t-\tau)$.

At each point in the parameter space grid, we applied an algorithm to calculate the periodicity of the solution. This algorithm involved tracking the number of rotations in the trajectory until it repeated itself, with a predefined threshold for the difference between $f(t)$ and $f(t+T)$ for multiple values of t, where T represented a candidate period duration. To be precise, the algorithm consists in the following steps and can be implemented in the desired programming language:

- A variable *rotationCounter* is initialized. At each point in the phase plane it grows by the value corresponding to the angular velocity measured over the vector field divided by 2π.
- For a given (x, y) time series (a trajectory), we obtain a time series for *rotationCounter*. Every integer *rotationCounter* (up to the desired maximum periodicity, in our case 8) in the time series becomes a candidate to be the n of the n-periodicity of the solution. T is the candidate period duration associated to that *rotationCounter*.
- Each of the *rotationCounter* candidates gets a score proportional to the average of the distances of $(x(t), y(t))$ and $(x(t+T), y(t+T))$ for s representative amount of times t.
- The minimum *rotationCounter* whose score is less than a fixed threshold is considered the n-period of the solution. That's the output of the algorithm.
- If no integer *rotationCounter* gets a score below the threshold, the algorithm outputs *null*.
- Precision can be increased by increasing the number of evaluated points in time to compute the distances and/or decreasing the threshold (it must be taken into account that decreasing too much the threshold without increasing accordingly the numerical precision of the integrator could produce undesired *null* outputs).

It must be noted that the classification is purely topological and not temporal. This means, for instance, that n-periodic solutions might have very different time periods for the same n. It also means that, for example, a 2-periodic solution may correspond to a longer time period than a 3-periodic one.

To visually represent our findings, we assigned integer values to classify the periodicity of the solutions at each explored grid coordinate of the parameter space. We used a color-coding scheme where unique colors were assigned to integer values up to 8. By superimposing the resulting areas from the four plots with transparency, we created a composite visualization shown in Fig. 5. White regions indicate cases where no periodicity was detected by the algorithm. These regions may correspond to higher period, quasiperiodic or chaotic solutions.

This approach allows us to gain insight into the presence and characteristics of periodic solutions across the $K \times \tau$ parameter space, providing a comprehensive view of the system's harmonic structure that emerges as a consequence of the applied

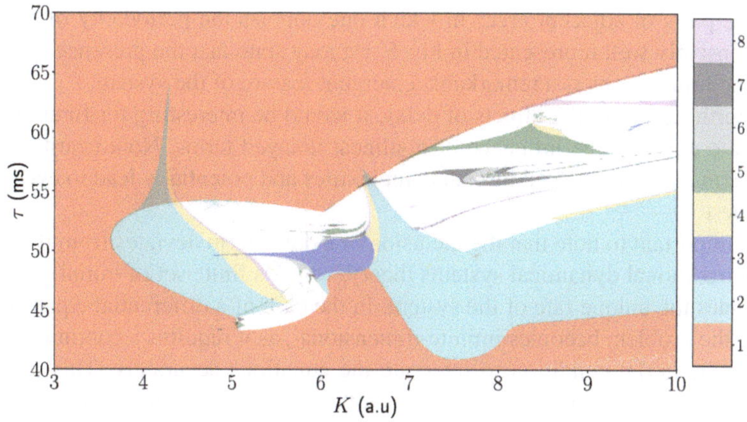

Fig. 5 Harmonic bifurcation map in delay-parameters space ($K \times \tau$) with $\rho_x = -2.3$, $\rho_y = -9$ and $\mu = 150$. Each colored region is linked to a periodic solution of n-periodicity, with n being an integer between 1 and 8. Transparency enables a way to visualize bistable zones since the plot was constructed by merging the areas of n-periodicity obtained by determining the period of the solutions of the system when integrated for the $K \times \tau$ parameter space explored in the four directions (increasing or decreasing each delay-related parameter). A rich spectrum of subharmonic solutions is observed. Uncolored regions are those where the solution was not n-periodic for any $n \leq 8$, i.e. higher periods, quasiperiodic or chaotic solutions

feedback. We observe a rich subharmonic structure, meaning that the periods of the oscillatory solutions of the system are integer multiples of a fundamental. Values below $K = 3$ and $\tau = 40$ ms are omitted in the plot because there are only period-1 solutions.

Figure 5 displays the regions where specific periodic solutions occur and where pairs of different n-periodic stable solutions coexist (so transparency allows to identify bistable regions as those with two colors). The color-coded map enables the identification of adjacent zones and their colors, allowing us to discern the permitted transitions between periodic solutions and their locations. This plot bears resemblance to Arnold's tongues observed in the study of the circle map [2].

The presence of multistability, where multiple periodic stable solutions coexist for certain parameter combinations, is a prominent feature observed in our results. This finding is consistent with previous research conducted on oscillator systems with time delay [7]. The identification and characterization of multistability adds to our understanding of the complex dynamics exhibited by the delayed version of the model.

The observed increase in complexity and the emergence of rich solutions in our model can be attributed solely to the introduction of a simple delay term, represented by a linear copy of the value of x at a past time $t - \tau$. No other modifications were made to the model. This highlights the profound impact of delayed feedback on the dynamics of the system.

Since pairs of adjacent areas in which one doubles the periodicity of the other are extensively well represented in Fig. 5, we may state that the presence of period-doubling bifurcations is a remarkable emergent feature of the system.

To further explore the effects of delay, it would be interesting for future investigations to consider the influence of nonlinear delayed terms. Nonlinearities in the delay term can introduce additional complexities and potentially lead to even richer dynamics.

It is important to note that the inclusion of a delay term deviates from the principles of traditional dynamical systems theory, where a finite set of initial conditions determines the unique fate of the system. In the case of a differential equation with delays, the problem becomes infinite-dimensional, as it requires a continuous range of states as initial condition to integrate the equations accurately. This departure from finite-dimensional dynamics adds another layer of complexity to the analysis and understanding of the system's behavior.

5 Discussion

The Wilson-Cowan model is widely recognized as a valuable tool for investigating the dynamics of neural populations. To gain a comprehensive understanding of this model, it is crucial to examine its behavior under various conditions. In this study, we specifically delved into the influence of time-delayed feedback on the dynamics of the Wilson-Cowan oscillator model. Our primary focus was on the emergence of a subharmonic structure in close proximity to a SNILC bifurcation. By exploring this aspect, we aimed to shed light on the intricate relationship between delayed feedback and the complex dynamics observed in neural systems.

The inclusion of the simplest feedback term in the Wilson-Cowan model revealed a complex subharmonic structure. It has been shown that multiple stable solutions exists for some parameter values. To which of the stable solutions the system locks is highly dependent of the previous states (and so, of the direction in which the parameter space is explored). This is a signature of hysteresis because the state of the system is dependent on its history: for two different vectors representing the states of the system between times $t - \tau$ and t, qualitatively different solutions may arise for exactly the same set of parameters.

In sensory physiology, phase locking refers to the firing of neurons preferentially at a certain phase of an amplitude-modulated stimulus. The neural oscillator we are modelling is one in which a linear delayed copy of the activity acts as a stimulus feeding back the system. The delay-time may be understood as the phase of a driving force to which the activity may synchronize either in a 1:1 way or as a subharmonic.

The implications of our findings extend beyond the realm of neuroscience and can be valuable in diverse fields such as physics and robotics. The complex dynamics observed in this model open up new possibilities for modeling and understanding intricate systems in various domains.

As a well-defined subharmonic structure is a characteristic pattern observed in the behavior of many biological systems, our results hold significant implications for both modeling and understanding the underlying mechanisms of these systems. By shedding light on the complex dynamics of the Wilson-Cowan model, our study may contribute to a better understanding of the behavior of biological systems and the potential applications of such models. Experiments to test the predictions could provide insights to the mechanisms that are actually involved in the production of periodical patterns with such characteristics.

Many animal behaviors are repeated multiple times over a relatively short period of time. It is likely that such repetitive actions are generated by oscillatory patterns of neural activity. In particular, birdsong is an animal model which is worth exploring under this perspective and has numerous parallels with human speech [3]. In many species, as canaries, the generation of the song involves the generation of periodic gestures, both respiratory and syringeal [9, 17]. The timing of these gestures is believed to be encoded in the HVC telencephalic nucleus. Research has provided evidence of glutamatergic and GABAergic synapses within the HVC, suggesting their involvement in the timing mechanisms of birdsong generation [10].

It is also known that HVC works as a relay center for both motor and auditory signals in the production and perception of birdsong [1]. It generates motor commands involved in song generation. Simultaneously, the HVC integrates auditory feedback by processing the auditory signals resulting from the bird's own vocalizations. The relevance of this mechanism varies depending on the degree of development of the HVC nucleus and the specific characteristics of each species.

Given the complex nature of birdsong and the involvement of motor and auditory systems, it is reasonable to speculate that some degree of processing delay exists. This delay can arise from various factors, including the time required for the song to be generated, propagated, and processed by the auditory system [5]. Studies have suggested that such delays may be involved in feedback mechanisms that regulate and fine-tune the production of birdsong [5].

Understanding the role of delayed feedback in the neural dynamics of birdsong can provide insights into how temporal processing and integration contribute to the generation and perception of complex vocalizations. Further experiments would be necessary to investigate the potential role of the subharmonic generation mechanisms described here in relation to the observed time durations of different syllable gestures.

In conclusion, this model not only holds significance within the realm of the neuroscience of vocal learning but also unveils a broader perspective on sensorimotor integration, a fundamental mechanism that resonates across diverse challenges in animal behavior. Our exploration into the modeling of this phenomenon, particularly in the context of two pivotal bifurcations, Hopf and SNILC, serves as a gateway to understanding a spectrum of periodic patterns. By establishing itself as a template, this study opens the door to a comprehensive investigation into sensorimotor integration, transcending specific examples and offering a lens through which the intricacies of this fundamental process can be unravelled in a more universal context.

References

1. Amador, A., Margoliash, D.: Auditory Memories and Feedback Processing for Vocal Learning, pp. 561–575. (2010)
2. Arnold, V.I.: Small denominators. I. Mapping the circle onto itself. Izv. Akad. Nauk SSSR Ser. Mat. **25**, 21–86 (1961)
3. Doupe, A.J., Kuhl, P.K.: Birdsong and human speech: common themes and mechanisms. Annu. Rev. Neurosci. **22**(1), 567–631 (1999)
4. Ermentrout, G.B., Kopell, N.: Parabolic bursting in an excitable system coupled with a slow oscillation. SIAM J. Appl. Math. **46**(2), 233–253 (1986)
5. Fukushima, M., Margoliash, D.: The effects of delayed auditory feedback revealed by bone conduction microphone in adult zebra finches. Sci. Rep. **5**, 03 (2015)
6. Hoppensteadt, F.C., Izhikevich, E.M.: Weakly Connected Neural Networks. Applied Mathematical Sciences. Springer, New York (2012)
7. Kim, S., Park, S.H., Ryu, C.S.: Multistability in coupled oscillator systems with time delay. Phys. Rev. Lett. **79**, 2911–2914 (1997)
8. Kuramoto, Y., Nishikawa, I.: Statistical macrodynamics of large dynamical systems. Case of a phase transition in oscillator communities. J. Stat. Phys. **49**(3), 569–605 (1987)
9. Long, M., Fee, M.: Using temperature to analyse temporal dynamics in the songbird motor pathway. Nature **456**, 189–194 (2008)
10. Mooney, R., Prather, J.: The HVC microcircuit: the synaptic basis for interactions between song motor and vocal plasticity pathways. J. Neurosci. Off. J. Soc. Neurosci. **25**, 1952–1964 (2005)
11. Prescott, S.A.: Excitability: Types I, II, and III, pp. 1–7. Springer New York, New York, NY (2013)
12. Reddy, D., Sen, A., Johnston, G.: Time delay induced death in coupled limit cycle oscillators. Phys. Rev. Lett. **80** (1998)
13. Ramana Reddy, D.V., Sen, A., Johnston, G.L.: Time delay effects on coupled limit cycle oscillators at hopf bifurcation. Phys. D Nonlinear Phenom. **129**(1), 15–34 (1999)
14. Ramana Reddy, D.V., Sen, A., Johnston, G.L.: Dynamics of a limit cycle oscillator under time delayed linear and nonlinear feedbacks. Phys. D Nonlinear Phenom. **144**(3), 335–357 (2000)
15. Schuster, H.G., Wagner, P.: A model for neuronal oscillations in the visual cortex. Biol. Cybern. **64**(1), 77–82 (1990)
16. Strogatz, S.H.: Nonlinear Dynamics and Chaos: With Applications to Physics, Biology, Chemistry and Engineering. Westview Press (2000)
17. Suthers, R.A., Margoliash, D.: Motor control of birdsong. Curr. Opin. Neurobiol. **12**(6), 684–690 (2002)
18. Wilson, H., Cowan, J.: A mathematical theory of the functional dynamics of cortical and thalamic nervous tissue. Kybernetik **13**, 55–80 (1973)
19. Winfree, A.T.: The Geometry of Biological Time. Interdisciplinary Applied Mathematics. Springer, New York (2001)
20. Wittenbach, J.D., Bouchard, K.E., Brainard, M.S., Jin, D.Z.: An adapting auditory-motor feedback loop can contribute to generating vocal repetition. PLOS Comput. Biol. **11**(10), 1–29 (2015)
21. Zeigler, H.P., Marler, P.: Neuroscience of Birdsong. Cambridge University Press (2008)

Unveiling the Intrinsic Dynamics of Biological and Artificial Neural Networks: From Criticality to Optimal Representations

Guillermo B. Morales, Serena di Santo, and Miguel A. Muñoz

Abstract Deciphering the underpinnings of the dynamical processes leading to information transmission, processing, and storing in the brain is a crucial challenge in neuroscience. An inspiring but speculative theoretical idea is that such dynamics, in order to entail a rich dynamical repertoire and optimize functional capabilities, should operate at the brink of a phase transition, i.e., at the edge between different collective phases. In recent years, research guided by the advent of high-throughput data and new theoretical developments has contributed to making a *quantitative* validation of such a hypothesis. Here we review recent advances in this field, stressing our contributions. In particular, we use data from thousands of individually recorded neurons in the mouse brain and tools such as a phenomenological renormalization group analysis, theory of disordered systems, and random matrix theory. These combined approaches provide strong evidence of quasi-universal scaling and near-critical behavior emerging in different brain regions. Moreover, we design artificial neural networks under the reservoir-computing paradigm and show that their internal dynamical states become near critical when we tune the networks for optimal performance. These results not only open new perspectives for understanding the ultimate principles guiding brain function but also towards the development of brain-inspired, neuromorphic computation.

G. B. Morales · S. di Santo · M. A. Muñoz (✉)
Departamento de Electromagnetismo y Física de la Materia and Instituto Carlos I de Física Teórica y Computacional, Universidad de Granada, Granada, Spain
e-mail: mamunoz@onsager.ugr.es

G. B. Morales
e-mail: guillermobm@onsager.ugr.es

S. di Santo
e-mail: serena@onsager.ugr.es

© The Author(s), under exclusive license to Springer Nature Switzerland AG 2025
M. Stich and J. Carballido-Landeira (eds.), *Nonlinear Dynamics for Biological Systems*, SEMA SIMAI Springer Series 40,
https://doi.org/10.1007/978-3-031-99044-1_2

1 Introduction: The Critical Hypothesis and Machine Learning at the Edge of Chaos

When neuroscientists poke into the electrophysiological processes of the brain—from microscopic individual neurons to large-scale whole-brain measurements—they systematically detect, even under conditions of quiet rest, an ongoing background of noisy and variable neural activity [1, 19, 20, 63, 68]. The fact that this ceaseless activity is energetically so costly, had researchers long wondering what crucial functionalities it might entail.

Diverse theoretical explanations for such a background state of reverberating activity have been proposed. A prominent one is that of the *balanced state or asynchronous state*, in which excitatory and inhibitory inputs to any given neuron typically cancel with each other, so that the mean input is below the activation threshold; however, local unbalances create a self-sustained fluctuation-driven noisy state, which may entail important functional advantages for information transmission, processing, etc. [9, 10, 37, 65, 82].

An alternative hypothesis considers that synaptic weights are tuned homeostatically, so that neuronal networks operate close to a *critical point*, i.e. at the edge between two different types of collective behavior [12, 14, 28, 29, 40, 44, 52, 55, 59, 62, 76, 84, 86]. In more generic terms, criticality emerges whenever the baseline phase loses its stability and shifts continuously to another phase; thus, it is also referred to as the *edge of instability* [42, 49, 50, 69]. Knowledge from Statistical Physics teaches us that near critical points there are long-range correlations, scale-invariant spatio-temporal patterns, enhanced collective responses, very rich dynamical repertoires, etc. [6, 52]. Therefore, it was conjectured that the brain, by being near criticality, could extract crucial advantages from such a plethora of spontaneously-generated collective properties [12, 66]. To be a bit more specific and make contact with the previous asynchronous-state scenario, one can consider a neural network composed of excitatory and inhibitory units; its overall state can be shifted (by varying, e.g., the intensity of the synaptic strengths) from the previously discussed asynchronous phase to a synchronous one, where collective oscillations emerge [36]. Thus, in particular, it has been argued that the brain could operate at the edge of a synchronization phase transition, so that it can jointly exploit the advantages of both asynchronous noisy states (e.g., for information processing) and synchronous ones (e.g., for coherent information transmission through oscillations) [34, 36, 61, 83, 88].

In parallel, advances in the area of artificial neural networks (ANNs) and machine learning have explored, since the seventies, the idea of computation at the "edge-of chaos" [4, 33, 56]: ANNs can exploit the combined advantages of stability (order) and responsiveness to inputs (disorder) when they operate at the borderline between order and "chaos", a regime that has been proven to be optimal from many different information-theoretic and computational perspectives [4, 8]. For example, it has been shown that among all *cellular automata*—i.e. discrete models of computation, composed by networks of binary units with local updating rules—only the ones posed at the edge of chaos, are capable of implementing computational algorithms [33]. Moreover, these ideas have guided the development of powerful and versatile machine-learning algorithms under the paradigm of "reservoir computing" [8, 31, 38, 41, 74].

In this chapter we will present a combined approach to the two previously described frameworks—biological and artificial neural networks—putting them under the common lens of the critical hypothesis.

The first and second section relies on high-throughput spiking data from the activity of thousands of simultaneously-recorded neurons over 16 different regions of the mouse brain [73]. A phenomenological renormalization-group approach then allows us to uncover strong evidence of (quasi-universal) scale-invariance across brain regions. Using diverse techniques based on random-matrix theory and theory of disordered systems, one can further ascribe such scale invariance to a close-to-critical dynamics, quantifying for each specific brain region its distance to the edge of instability.

In the third section we follow the seminal work of Stringer et al. [75], showing how recorded patterns of activity in the mouse visual cortex generate internal representations of external stimuli in an optimal way. To find whether such optimal representations are linked to an underlying critical dynamics, we construct Echo State Networks [31, 39] and train them to classify images. Remarkably, only when the network dynamics is tuned close to the edge of instability we recover the same encoding properties observed in the mouse visual cortex, concomitant with maximal classification performance.

All together the results reviewed here provide us with a general standpoint to analyze neural networks under the lenses of criticality and computation at the edge of chaos. In particular, our analyses reveal that close-to-critical behavior, which is found to be ubiquitous across the brain, can also allow for the construction of optimal input representations in both, biological and artificial neural networks.

2 Resting-State Brain Activity Shows Scale Invariance

In this section, we show how statistical properties of spontaneous neural activity exhibit scale invariance as well as striking similarities across different brain regions in the mouse brain, despite their inherent heterogeneity. The following analyses rely on the phenomenological renormalization group approach formerly devised by Meshulam et al. in [47]. For the sake of simplicity, we skip here some technical aspects; the interested reader can find all the details in [50].

2.1 Introducing a Phenomenological Renormalization Group

Testing whether neural networks operate near a phase transition is a daunting task, not devoid of controversy within the scientific community [3, 80]. As mentioned in the Introduction, empirical support for the criticality hypothesis has typically relied on the observation of scale-free bursts of activity, called "neuronal avalanches" [2, 27], but other approaches exist, including the study of long-range spatio-temporal correlations [57, 61]; the search for statistical critical-like patterns of activity [79]; or the analysis of whole-brain models fitted to match empirically observed correlations [11] (see e.g. [14, 45, 52, 55, 60, 86] for recent reviews).

A concept that is intimately linked with that of criticality is the existence of scale-invariance. At criticality, a system exhibits fractal-like scale-invariance, meaning that its properties remain unchanged under rescaling, so that the behavior is characterized by universal features that are independent of the specific details or size of the system [52]. This is often captured by power-law distributions—i.e. the mathematical fingerprint of scale invariance [71]—of diverse quantities.

The presence of scale-invariance or *scaling* at critical points can be understood because of the divergence of the system's correlation length at such points, implying that—as criticality is approached—the system becomes correlated over larger and larger scales. Notice that the existence of long-range spatio-temporal correlations gives rise to cooperative effects, where fluctuations at one scale influence fluctuations at coarser scales; in other words there is no privileged scale and the system is thus scale invariant.

To analyze this type of problems with many scales, the Renormalization Group (RG) emerges as a mathematical tool, widely used in the realm of Statistical Physics, that allows one to study the behavior of a system as we zoom in or out, capturing the universal features that persist across different scales [6, 22, 85]. By exploring the properties of the underlying microscopic constituents, the RG enables us to uncover the emergence of scale invariance.

Typically, an RG analysis involves the construction of effective descriptions of a microscopic system at progressively larger spatial scales. In the original real-space RG, as proposed by Kadanoff for spin-systems, one has to "coarse-grain" neighboring spins into blocks, which are then clustered into new blocks at successive steps of the RG transformation [22]. The problem, when one tries to adapt this methodology to neural data, is that one typically does not know which neurons are "neighbors" of which, as one usually lacks information about the detailed connectivity patterns at such microscopic levels. Meshulam et al. found a workaround to this problem in what they termed a "phenomenological Renormalization Group" (PRG) approach; their approach led them to find strong evidence of scale-invariant properties in the recorded neural activity of the CA1 region of the mouse hippocampus [47].

To understand the context and results of this study, let us consider a set of N neurons, such that the empirically-measured activity of the i-th neuron at a specific time t_j is represented as $x_i(t_j)$, with $j \in [1, T]$, describing a discrete set of non-overlapping time bins up to a maximum time T. The coarse-graining procedure then proceeds as follows: instead of grouping neighboring neurons using a criterion of spatial vicinity, one clusters them based on a criterion of maximal pairwise Pearson's correlation:

$$C_{ij} = \frac{\langle \delta x_i \delta x_j \rangle}{\sqrt{\langle (\delta x_i)^2 \rangle \langle (\delta x_j)^2 \rangle}}, \tag{1}$$

where averages are computed across the available discrete time steps and δx is the deviation from the mean. More specifically, at the k-th coarse-graining step, one selects the two most correlated block-neurons, i and j_{*i}, and combine their activities into a new "block-neuron" coarse-grained variable given by:

$$x_i^{(k)} = z_i^{(k)} \left(x_i^{(k-1)} + x_{j_{*i}}^{(k-1)} \right), \tag{2}$$

where $k = \{1, ..., n_{steps}\}$ indexes the RG step—one takes $x_i^{(0)}$ to be the original timeseries of the i-th neuron as extracted from the data—and the normalization factor, $z_i^{(k)}$, is chosen such that the average non-zero activity of the new variables $x_i^{(k)}$ is equal to one. The procedure then is iterated by considering the second most correlated pair of neurons, and so on, until a set of $N_k = N/2^k$ coarse-grained "block-neurons" is obtained, with each block containing the summed activity of $K = 2^k$ original neurons (see Fig. 1 for a schematic representation).

Once the coarse-graining procedure has been completed, the goal is to study how the statistical properties of the system change as one moves progressively across scales. Following [47], here we focus on four quantities that have a clear interpretation in the classical theory of critical phenomena [6]:

1. The variance of the non-normalized coarse-grained variables.

$$M_2(K) = \frac{1}{N_k} \sum_{i=1}^{N_k} \left[\left\langle \left(\sigma_i^{(k)} \right)^2 \right\rangle - \left\langle \left(\sigma_i^{(k)} \right) \right\rangle^2 \right] \tag{3}$$

where $\sigma_i^{(k)}$ is the summed activity of the original variables inside a given cluster and N_K is the number of clusters at scale k. Notice that, for totally independent random variables, one would expect the variance to grow linearly in K (i.e., $M_2(K) \propto K$), whereas if variables were perfectly correlated $M_2(K) \propto K^2$. Non-trivial scaling is therefore characterized by

$$M_2(K) \propto K^\alpha \tag{4}$$

with a certain intermediate value of the exponent $1 < \alpha < 2$.

2. The "free-energy" for the coarse-grained variables, defined as:

$$F(K) = -\log(S_K). \tag{5}$$

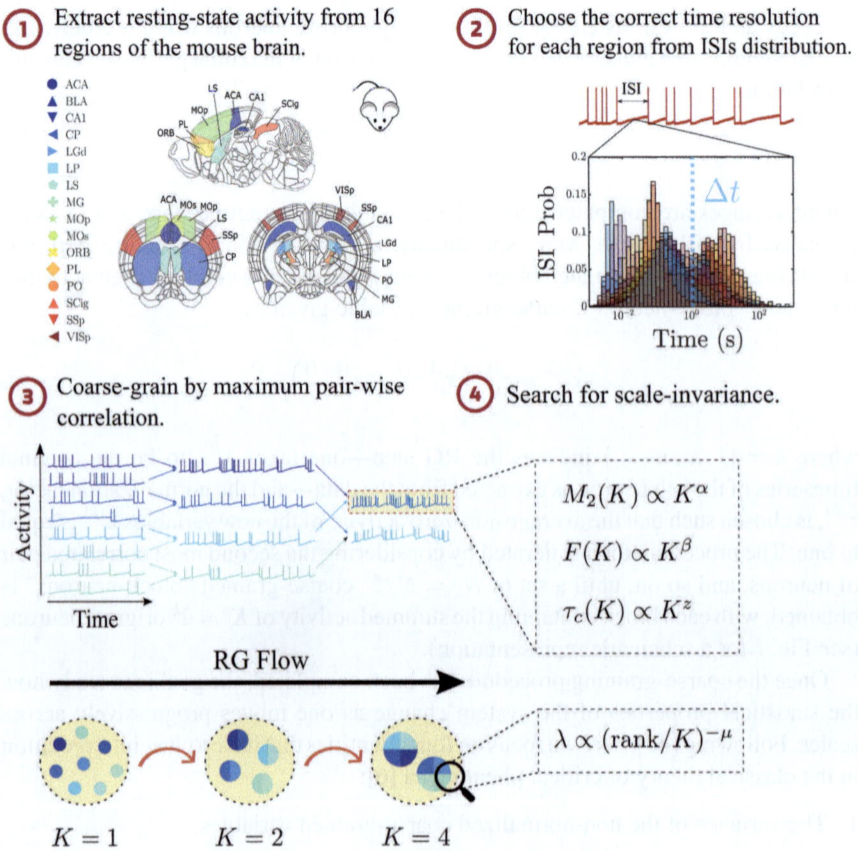

Fig. 1 Workflow of the phenomenological renormalization group analysis. (**1**) The raw data is filtered for resting-state activity in regions with at least $N = 128$ recorded neurons. (**2**) For each region, activity is discretized in bins of width Δt equal to the geometric mean of the inter-spike-interval (ISI) distribution. (**3**) The original system is progressively coarse-grained by clustering together pairs of most-correlated neurons, so that at step k each new variable contains the summed activity of $K = 2^k$ original neurons. (**4**) By changing the scale K, different properties are analyzed as we move towards more coarse-grained descriptions of the system

where, S_K is the probability that a given coarse-grained neuron is silent at any time step. As more and more of the initial variables σ_i are grouped into cluster variables $x_i^{(k)}$, one would expect that the probability of having "silent" block-neurons (i.e., the probability that all neurons inside a cluster are silent) decreases rapidly with the size K of the clusters, leading to:

$$F(K) \propto K^{\beta}, \tag{6}$$

where, in particular, one expects an exponential decay, i.e. $\beta = 1$, for initially independent variables [47].

3. The autocorrelation function of the coarse-grained variables. Given that, commonly, fluctuations at larger spatial scales relax with a slower characteristic time scale, one should expect the autocorrelation function of the coarse-grained variables to decay more slowly as one averages over more neurons. In particular, for the k-th step of the RG flow, one has:

$$C^{(k)}(t) = \frac{1}{N_k} \sum_{i=1}^{N_k} \frac{\langle x_i^{(k)}(t_0) x_i^{(k)}(t_0 + t) \rangle - \langle x_i^{(k)} \rangle^2}{\langle (x_i^{(k)})^2 \rangle - \langle x_i^{(k)} \rangle^2} \tag{7}$$

which, at the steady state, is independent of t_0. Assuming that correlations decay exponentially in time with a characteristic time scale $\tau_c^{(k)}$ (i.e., $C^{(k)}(t) = e^{-t/\tau_c^{(k)}}$) at each coarse-graining level, dynamical scaling implies that the average correlation function collapses into a single curve when time is re-scaled by the characteristic time scale: $C^{(k)}(t) = C(t/\tau_c^{(k)})$ and that these characteristic times obey scaling with the cluster size:

$$\tau_c(K) \propto K^z, \tag{8}$$

where z is the dynamical scaling exponent.

4. The spectrum of the cluster covariance matrix. Finally, as argued in [47], if correlations are self-similar across scales, then we should see this by looking inside the clusters of size K. In particular, the eigenvalues of the covariance matrix must obey a power-law dependence on their fractional rank:

$$\lambda \propto \left(\frac{rank}{K} \right)^{-\mu}. \tag{9}$$

Thus, the exponents α, β, z, and μ characterize the possible scale-invariant behavior of the system under study.

2.2 RG Scaling Exponents Are Quasi-universal Across Brain Regions

By applying the previously introduced PRG to neural recordings (including more than one thousand neurons) of the mouse hippocampus, Meshulam et al. observed the emergence of a non-Gaussian fixed point for the non-zero activity distribution; non-trivial scaling of the variance, "free-energy" and autocorrelation times; as well as spatial scaling reflected in the power-law dependence with the fractional rank for the eigenspectrum of the covariance inside clusters [47]. These pieces of evidence reveal that the underlying neural dynamics is scale invariant (see below for a more detailed explanation).

To analyze the possible robustness or universality of these findings, we decided to analyze different brain regions in the mouse brain. In particular, most of the forthcoming analyses rely on the empirical electrophysiological data presented by Steinmetz et al. in [73], where the activities of thousands of individuals neurons were simultaneously recorded at a high temporal resolution (200 Hz) in several mouse brain regions (see Fig. 1). To conduct a comprehensive analysis, we initially segregated the original data into "resting-state" and "task-related" activity, focusing on the former for the purpose of our study (Fig. 1(1)).

Before delving any further into the results, let us briefly mention one important aspect that needs to be (and it is rather often not) considered when applying PRG to experimental data: time binning. Typically, studies involving neural recordings of activity, both at the single-neuron level and at lower spatial resolution scales (EEG, fMRI, etc.), involve the discretization of the time series into time bins of a certain width Δt, so that the activity $x_i(t_j)$ of neuron i at bin j is defined as $x_i(t_j) = \int_{t_j}^{t_j+\Delta t} x_i(t)dt$. Whether the time bin size is chosen to match the time resolution of the recording technique, a relevant timescale of the input task, or it is just arbitrarily selected, the truth is that correlations and correlations-based properties of the system can depend drastically on the choice of Δt. This is especially true for single-neuron activity, given that neurons within the same population can operate at broadly different time-scales, with spiking frequencies ranging from milliseconds to seconds (see Figs. 1A, S4, and S5 in [50]). To cope with such heterogeneity, in the forthcoming analyses, the relevant timescale for each brain region (and thus our choice of Δt) is defined as the *geometric mean* of the inter-spike-interval distribution (see Fig. 1(2)). As a sanity check, we observed that this choice of Δt reproduces to great extent the previously reported scaling exponents found in [47] for the mouse hippocampus, using Steinmetz et al.'s recordings for this very same area. This result is not trivial and supports the robustness of the observe scale invariance, as the data and recording techniques were very different in both studies.

Figure 2 summarizes the analysis presented in [50] for the emergence of scale-invariant properties with the RG flow in the recorded activity. In particular, the top left panel, shows an almost perfect scaling of the variance of the coarse-grained variables with an average exponent $\overline{\alpha} = 1.38 \pm 0.08$ across regions, which is in between the expected one for uncorrelated ($\alpha = 1$) and fully-correlated variables ($\alpha = 2$). On the other hand, the "free-energy" as defined in Eq. (5), also exhibits a clear scaling with the cluster size (Fig. 2, top right panel), with an average exponent $\overline{\beta} = 0.79 \pm 0.03$ across regions. Finally, temporal and spatial scaling are manifested in the bottom panels of Fig. 2, respectively, with the former showing the observed dynamical scaling in the decay of the coarse-grained neuron's autocorrelation times $\tau_c(K)$, with an average exponent $\overline{z} = 0.22 \pm 0.05$ across regions. For clarity, the power-law decay of the covariance spectrum is plotted only for one of the regions (ACA), where the fractional-rank ($rank/K$) dependence is manifested in the collapse of the curves at different levels of coarse graining.

To ensure the consistency of the results, we have additionally confirmed that the documented exponent values show minimal variations when the time-discretization

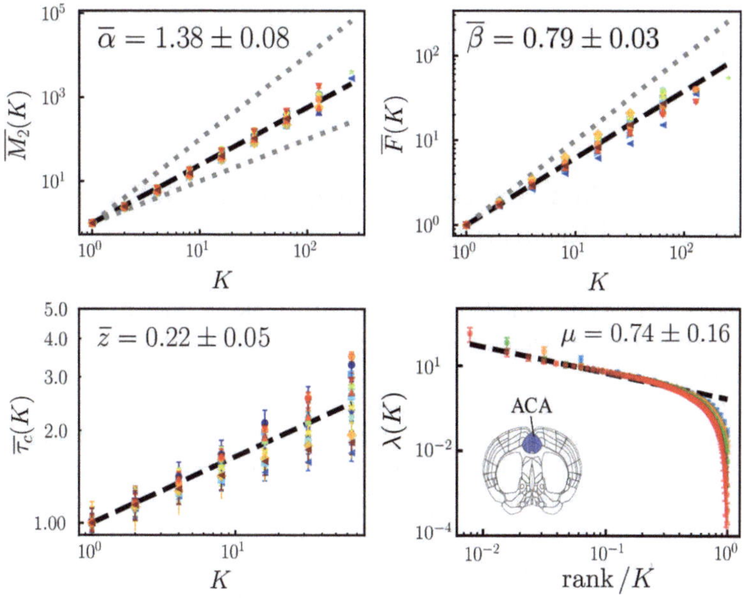

Fig. 2 Phenomenological RG analyses unveils the existence of quasi-universal scaling over 16 different regions of the mouse brain. Top left panel: Variance of the non-normalized activity as a function of the block-neuron size, K, in double logarithmic scale (upper and bottom dotted lines, with slopes 2 and 1, mark the fully correlated and independent limit cases, respectively). Top right panel: Scaling of the free energy F_k, as defined in Eq. (5) (the dotted line corresponds to the expected behavior for un-correlated variables). Bottom left panel: Scaling of the characteristic correlation time τ_c as a function of K in double logarithmic scale for the different areas. Bottom right panel: Scaling of the covariance matrix spectrum for the resting-state activity in clusters of size $K = \{16, 32, 64, 128\}$ (blue, yellow, green and red markers, respectively) for one of the considered brain regions. The rank-ordered eigenvalues decay as a power law of the fractional rank (rank/K) and, even more remarkably, the curves and in particular the cut-offs at different K collapse into the same curve. To facilitate the comparison between areas with different number of neurons, the points of the curves in the first three subplots have been normalized by the value of each quantity at $K = 1$ (e.g., $\overline{M_2}(K) = M_2(K)/M_2(K = 1)$). Errorbars are computed as the standard deviation across split-quarters of data, with lengths typically smaller than the marker size. Adapted with modifications from [50]

bin is changed, with the exception of μ, which displays an increase during longer time-scales beyond the population activity's typical inter-spike-interval (see Fig. S6 in [50] for a more in depth discussion).

At the light of the above results, one can confidently state that strong signatures of scale-invariance with quasi-universal exponents are indeed observed across brain regions.

Does such a scale invariance naturally follow from an underlying critical dynamics? The answer to this question is not straightforward as, e.g., an alternative explanation to the observed scale-invariance has been recently derived in terms of latent fields in a model with no critical point in its dynamics [51, 53]. Thus, in the next section we tackle this question not only by looking for signatures of scale invariance, but actually inferring from the data how far the dynamics lies from a critical point.

3 Estimating the Distance to the Edge-of-Instability from Neural Recordings

Given an empirical system showing some degree of scale invariance, as the previously discussed ones, can we quantify how close it is to the edge of a phase transition or critical point?

The logic flow we pursue is as follows: Let us (i) choose the simplest neuronal model exhibiting a phase transition between two different regimes; (ii) introduce two different methods to infer from them the distance to criticality from neural activity measurements; (iii) apply and compare the two methods to the recordings in the different regions of the mouse brain analysed above [73].

In what follows we employ a simplified linear firing-rate neural model. This should not be taken literally as a model for actual coupled neurons but, rather, as an effective model that has been proven to work remarkably well as a linear-response approximation of much more complex systems including ingredients such as spiking neurons, non-linear interactions, etc. [17, 30].

In particular, we consider a *linear-rate model* [30, 70], in which the time evolution of N linearly interacting units, each one described by a continuous variable $x_i(t)$ representing its activity or "firing rate" at instant t and connected with each other through a synaptic-connectivity matrix J_{ij}:

$$\dot{x}_i(t) = -x_i(t) + \sum_{j=1}^{N} J_{ij} x_j(t) + \xi_i(t). \tag{10}$$

where $\xi_i(t)$ are noisy external inputs that, for simplicity, we assume to be zero-mean uncorrelated Gaussian white noises with $\langle \xi_i(t)\xi_j(t')\rangle = \delta_{ij}\delta(t-t')$. The synaptic connectivity matrix is assumed to be random, i.e., its elements J_{ij} are sampled from a Gaussian distribution with zero mean and variance g^2/N (i.e., $J_{ij} \sim \mathcal{N}(0, g^2/N)$).[1]

A well-known result from random-matrix theory, the circular law, asserts that for $N \to \infty$ a random matrix with Gaussian-distributed entries has a spectrum of eigenvalues distributed uniformly over a disc in the complex plane of radius $\lambda_{max} = g$ [24, 25, 46, 78] (see Fig. 3). Thus, denoting by I the identity matrix, the steady-state

[1] Let us note that the connectivity pattern described by J_{ij} is not realistic from a biological point of view. In particular real neural networks obey Dale's law, meaning that each neuron either excites or inhibits its postsynaptic neurons, which implies that all elements along a column of the connectivity matrix of real neurons should have the same sign. The spectrum of such matrices does not obey the circular law [64], nevertheless, the results presented below generalize easily to excitatory-inhibitory connectivity matrices [17, 30].

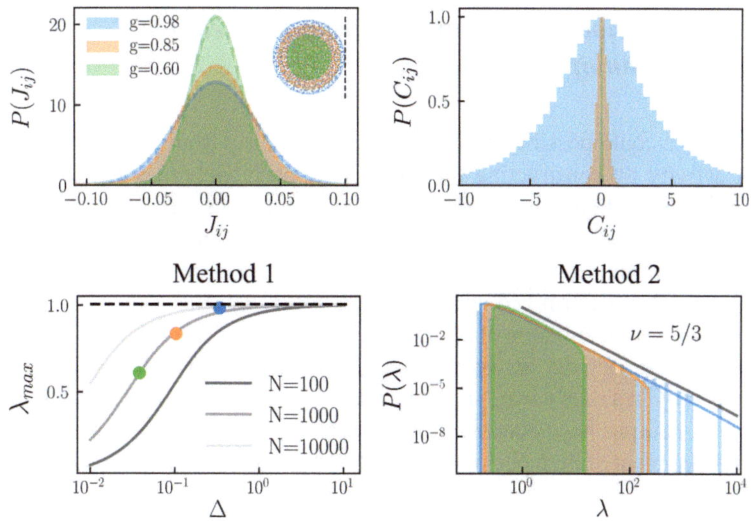

Fig. 3 Statistics of pairwise covariances provide two different measures of distance to the edge of instability. For each panel, we compare simulated results for the linear-rate model using 3 different values of g. Top left panel: The entries of the connectivity matrix J_{ij} in the linear-rate model follow a Gaussian distribution of zero mean and variance g^2/N. The spectrum of the matrix is contained in a circle of radius g in the complex plane (inset), with $\lambda_{max} = 1$ marking the edge of instability (black dashed line). Top right panel: entries of the sampled spike-count covariance matrix C_{ij} for $N = 10^3$. Histograms are re-scaled to their maximum value for visualization purposes. Bottom left panel: maximum eigenvalue of the connectivity matrix J as a function of the relative width of the distribution for the covariance matrix entries. Full lines are drawn using Eq. (14) for different values of N, whereas circles are obtained through simulations of the linear-rate model for the 3 values of g considered (black dashed line represents the edge of instability). Bottom right panel: Distribution of the eigenvalues of the covariance matrix in the linear-rate model. Full lines are plotted using the analytical expression in Eq. (15), whereas histograms are obtained from simulations of the linear-rate model. As $g \to 1$ the tail of the distribution approaches a power law with exponent $\nu = 5/3$ (black continuous line)

of Eq. (10) is stable if all the eigenvalues of the matrix $(-I + J)$ are negative, i.e., if $\lambda_{max} = g < 1$. In other words, the distance to the edge of instability is fully characterized by the maximum eigenvalue of the connectivity matrix.

Therefore, having access to the full synaptic connectivity matrix in the experimental setup one could automatically have an estimate of the distance to the edge of instability (under the assumption of the underlying linear dynamics). Since inferring such a full connectivity matrix from actual data is an extremely daunting task from the experimental point of view [26, 87], one needs to resort to alternative strategies, as we will see in the following sections.

3.1 First Method: An Estimation from the Relative Dispersion of Covariances

Typically, one could possibly infer a connectivity matrix from experimental recordings of activity by constructing maximum entropy models that aim to match the observed mean and variance of the empirical correlations (see, e.g., [15] and refs. therein). However, this would imply estimating the N^2 entries of the connectivity matrix when in fact, as we saw, the dynamical stability in this simple linear models can be completely determined with just its largest eigenvalue.

Remarkably, an estimation of λ_{max} can be obtained from a measure of the pairwise covariance of the activity across neurons. This quantity is much easier to measure experimentally in a robust way, since simultaneous recordings of the activity of large numbers of neurons have become more and more attainable in recent years, thanks to the development of high performance neural probes, as we have already discussed [21, 73, 89].

To estimate λ_{max}, one needs to compute the so-called spike-count covariance matrix (sometimes also called noise covariance or long-time-window covariance). The elements of this matrix are pairwise covariances of the time-integrated activity across many samples, i.e.:

$$C_{ij} = \lim_{\Delta t \to \infty} \frac{1}{\Delta t} \langle \Delta s_i(t) \Delta s_j(t) \rangle, \tag{11}$$

where averages are taken over samples and $\Delta s_i(t)$ is computed as:

$$\Delta s_i(t) = \int_t^{t+\Delta t} (x_i(t') - \langle x_i \rangle) dt'. \tag{12}$$

Notably, this covariance matrix is linked in a rather model-independent way to the connectivity J through the following relationship [17, 30, 58, 81]:

$$C = (I - J)^{-1}(I - J^T)^{-1} \tag{13}$$

with the main assumption that inputs are uncorrelated. A simple proof of this equation is presented, for the sake of completeness, in Appendix 1.

Note, however, that one cannot calculate J by inverting Eq. (13) for an empirically measured C matrix, because the recorded neurons represent only a small subsample of the whole local network so that the inversion strategy is plagued with pitfalls. To circumvent this difficulty, Dahmen et al. recently devised a method to estimate the maximum eigenvalue λ_{max} of the connectivity matrix J based on the distribution of the entries of C [17]. More specifically, using ideas and tools from the physics of disordered systems, such as spin glasses, they found the following relationship:

$$\lambda_{max} = \sqrt{1 - \sqrt{\frac{1}{1 + N\Delta^2}}}, \tag{14}$$

where $\Delta = \delta c / \bar{c}$, being δc the dispersion of the entries of the spike-count covariance matrix, calculated as the standard deviation of the out-of-diagonal terms of C, and $\bar{c}$ the mean variance, i.e. the mean of the diagonal terms.

Thus, this rather elegant result provides us with an overall estimator of the network stability which depends only on the relative dispersion of empirically measured pairwise covariance values and the system size. Moreover, Dahmen et al. show that this result is insensitive to the details of the underlying dynamics and connectivity; for instance they show that Eq. (14) still holds (as a linear-response approximation) when considering spiking neurons dynamics or excitatory-inhibitory connectivities (see [17, 54] details).

As we will see in the following section, an alternative estimation of λ_{max} can be obtained using the full eigenspectrum of the empirical covariance matrix.

3.2 *Second Method: An Estimation from the Eigenvalue Spectrum of Covariances*

In a mathematical tour-de-force, Hu and Sompolinsky took the theory of linear-rate models one step further, deriving an analytical form for the full probability density of the long-time window covariance eigenvalues in the system described by Eq. (10) [30].

$$p(\lambda) = \frac{3^{\frac{1}{6}}}{2\pi g^2 \lambda^2} \left[\sum_{\xi=1,-1} \xi \left((1 + \frac{g^2}{2})\lambda - \frac{1}{9} + \xi \sqrt{\frac{(1-g^2)^3 \lambda (\lambda_+ - \lambda)(\lambda - \lambda_-)}{3}} \right)^{\frac{1}{3}} \right], \tag{15}$$

for $\lambda_- \le \lambda \le \lambda_+$, with

$$\lambda_{\pm} = \frac{2 + 5g^2 - \frac{g^4}{4} \pm \frac{1}{4} g (8 + g^2)^{\frac{3}{2}}}{2(1-g^2)^3}, \tag{16}$$

while $p(\lambda) = 0$ for values of λ out of the support, i.e. for $\lambda > \lambda_+$ and $\lambda < \lambda_-$. Observe, in particular, that from Eq. (16) the upper limit of the support λ_+ diverges in the limit $g \to 1$, i.e. close to the edge-of-instability.

In fact, in such a limit, the distribution $p(\lambda)$ develops a long (power-law) tail of large eigenvalues:

$$\lim_{g \to 1} p(\lambda) \sim \frac{\sqrt{3}}{2\pi} \lambda^{-\frac{5}{3}}. \tag{17}$$

Figure 3 shows that the above analytical results, which were derived in the limit of $N \to \infty$, match accurately the spectrum obtained numerically in simulations of the linear-rate model with $N = 10^3$ units. Thus, in practice, provided one has simultaneously recorded enough neurons (on the order of hundreds) and has enough samples as to meaningfully compute their covariance matrix, it is possible to infer the dynamical regime of the recorded population simply by fitting the sampled covariance matrix eigenvalue spectrum to the theoretical distribution given by Eq. (15), using g as the fitting parameter. This will constitute our second method to estimate the distance to the edge-of-instability from data.

We remark that, although the above expression for the density of eigenvalues was derived assuming a linear-rate model of recurrently connected neurons, Hu and Sompolinsky showed that it provides also an excellent fit to the numerical spectrum of a network of nonlinear rate neurons driven by external noise. Let us caution, in any case, that nonlinearities may introduce a bias in the inferred g value (see [30] for more details).

3.3 *Mouse Resting-State Activity Shows Close-to-Critical Dynamics*

In this section, we apply the two different methods of estimating λ_{max}, as discussed above, to the previously introduced dataset of Steinmetz et al. [73]. To avoid confusion, we will denote by $\hat{g}_d$ and $\hat{g}_s$ the empirical estimates of the distance to criticality obtained using the actual data from the relative dispersion of covariances (first method [17]) and their eigenvalue spectrum (second method [30]), respectively.

First of all, since both methods rely on the computation of the long-time window covariance matrix, Eq. (11), we first need to split the original time series of spiking activity for each neuron into T samples of width Δt. In practice, the condition $\Delta t \to \infty$ can be approximated by choosing a window large enough for autocorrelations to decay (Fig. 4(1)). In the following analyses, a characteristic time τ for each region was obtained from an exponential fit to the mean autocorrelations, then a timebin $\Delta t = 1s$ was chosen so that $\Delta t > \tau$ while maximizing the number of samples available. Next we calculate the spike-counts for each neuron by integrating the activity within the previously defined time windows and compute the covariance matrix across neurons (the distribution of entries of such a matrix is shown in Fig. 4(2) for one of the regions).

A first estimate of the distance to criticality in each region is obtained using Eq. (14) (see Fig. 4(3)). By taking random subsamples of the recorded neurons, one can generate a curve of the estimated maximal eigenvalue $\hat{g}_d$ as a function of network size. Then, to make a sensible comparison between regions with different number of recorded neurons, we extrapolate the value of $\hat{g}_d$ for a common number of neurons $N = 10^4$ in each area (Fig. 4(3), left). As we can observe in Fig. 4(3) (right), all considered areas turn out to be very close to the edge of instability.

Alternatively, one can infer g from the best-fitting parameter $\hat{g}_s$ minimizing the Cramer von-Mises distance between the empirically obtained cumulative distribution and the theoretically-derived one $F(\lambda) = \int_{-\infty}^{\lambda} p(\lambda)d\lambda$ [30]. To illustrate this, in Fig. 4(4) we plot the empirical distribution of covariance eigenvalues for an example region, together with the best fitting distribution $p(\lambda)$ (obtained for $\hat{g}_s = 0.95$, very close to the critical value). Moreover, we report for comparison the best-fitting Marchenko-Pastur eigenvalue distribution (i.e., the one expected for uncorrelated random variables), which clearly fails to capture the long tail of the empirical eigenvalue distribution.

To summarize, Fig. 4(4) presents a comparison of the two alternative estimates of the distance to criticality, $\hat{g}_d$ and $\hat{g}_s$, for all the 16 considered regions of the mouse brain, revealing that—within errorbars—they are typically in excellent agreement. Thus, the above results let us conclude that all the analyzed regions are, to greater or lesser extent, close to the edge of instability. Understanding the origin and meaning of the existent variability across regions is beyond our scope here.

3.4 Closing the Loop: Do Critical Models Show Scale Invariance?

In previous sections we have described a PRG scheme that can be used to reveal genuine scale invariance in a dataset of simultaneous neuronal recordings (Sect. 2.1), as well as two different ways to infer how close or far to the edge of instability are these systems, both based on measures of the covariance of the units' activity (Sect. 3). The missing link is that, thus far, we do not have any evidence that systems—in particular theoretical models—poised near the edge of instability give rise to scaling behavior compatible with the one revealed using the PRG analysis in actual neural data.

Therefore, in order to complete the picture, here we run simulations on a linear-rate model of randomly connected neurons that follow Eq. (10), for which we can control the distance to the edge of instability through the coupling parameter g, creating a synthetic dataset over which to apply the previously introduced PRG procedure. We a priori expect to find non-trivial scaling for $g \simeq 1$ but not for $g \ll 1$ and, moreover, we wonder whether the scaling exponents, if they can be properly measured, are similar to empirical ones.

The results of these analyses are illustrated in Fig. 5; in particular they reveal that:

- The variance of the activity scales with cluster size with a trivial exponent $\alpha \simeq 1$ for $g = 0.2$ and $g = 0.6$, expected for independent units; however, for $g = 0.95$,—sufficiently close to the limit of instability—it shows the non-trivial exponent $\alpha = 1.38$, in surprisingly good agreement with the average exponent measured for the spiking neurons ($\bar{\alpha} = 1.38 \pm 0.08$, Fig. 2, top right panel).
- The characteristic time for the autocorrelation function shows a disruption of scaling for $g = 0.2$ and $g = 0.6$ for large cluster sizes and a close-to-trivial scaling for small cluster sizes; however, for $g = 0.95$ the scaling holds for the all the

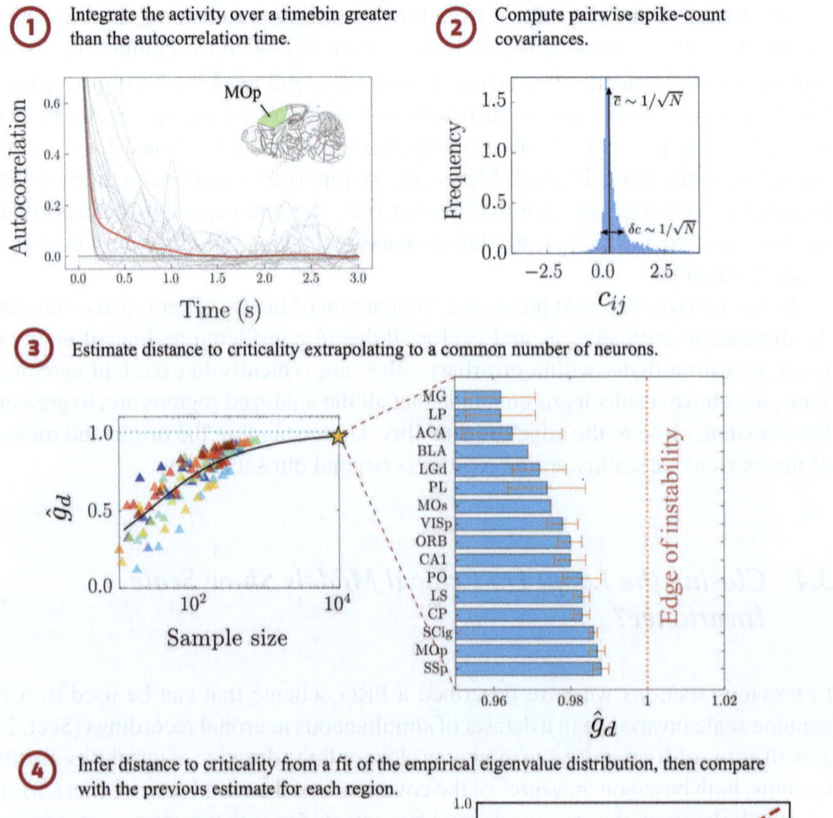

Fig. 4 Distance-to-criticality analysis locates the activity in 16 regions of the brain near the edge-of-instability. (**1**) Decay of autocorrelations in the MOp region (grey lines: 30 randomly chosen neurons; red line: average over neurons). (**2**) Distribution of pairwise covariances in the MOp region, showing a strong peak around 0. (**3**) Estimated distance to criticality as a function of the number of neurons N in the sample. For each region (color coded) the points were fitted to Eq. (14) to obtain an empirical estimate $\hat{g}_d$ of λ_{max} at a common number of neurons $N = 10^4$. Only an average over such fitted curves (black line) is shown for visualization purposes. (**4**) Left: Covariance eigenvalues distribution for the MOp region, together with the best-fitting Marchenko-Pastur distribution (green line) and the best-fitting eigenvalue distribution for a linear-rate model of randomly connected neurons (red line, $\hat{g}_s = 0.95$). The inset shows a close-up of the corresponding fits in a log-log scale. Right: Estimated values of $\hat{g}_d$ (dispersion method, [17]) and $\hat{g}_s$ (spectral method, ([30]) for each of the 16 regions. In all cases, errors are computed as the SD over different recordings of the same region. Adapted with modifications from [50]

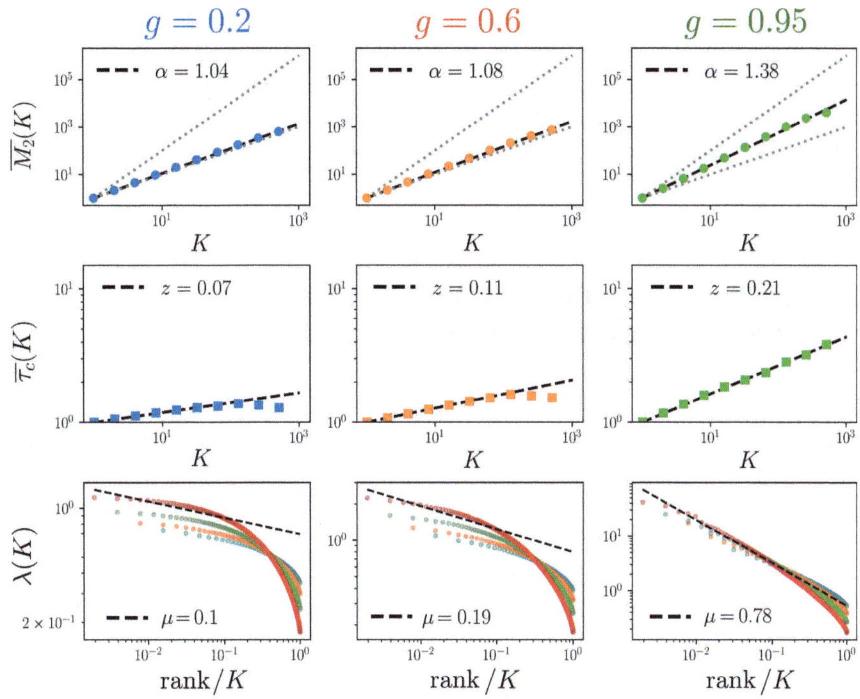

Fig. 5 A phenomenological RG analysis for the linear-rate model shows signatures of scale-invariance when the system is close to the edge-of-instability. RG over a randomly recurrent network of linear-rate neurons for different values of the overall coupling strength g. Top: Variance of the non-normalized activity, Middle: scaling of the characteristic correlation time, Bottom: Covariance matrix spectrum for the activity in clusters of size K, as a function of the block-neuron size. The system closest to the edge-of-instability ($g = 0.95$) shows scalings and power-law exponents similar to the ones observed in real data. The method also detects that further away from the edge-of-instability the model lacks of non-genuine scale invariance: (i) the variance of activity shows a scaling exponent compatible with the one for uncorrelated variables; (ii) the rank-ordered spectrum is close-to-flat and shows a cutoff for large cluster sizes, and (iii) the collapse across RG steps of the rank-ordered covariance matrix spectrum is lost. Parameter values: $N = 1024$, $\sigma = 1$, $\tau = 1$. Adapted with modifications from [50]

considered cluster sizes and the measured dynamical scaling exponent is very close to average experimental value ($\bar{z} = 0.22 \pm 0.05$, Fig. 2, bottom left panel).

- The rank-ordering of the spectrum of eigenvalues of the covariance matrix has a cut-off that changes with cluster-size quite dramatically for $g = 0.2$ and $g = 0.6$. On the contrary, for $g = 0.95$ there is a—not perfect but much better—collapse of the curves at different levels of coarse-graining and a power-law trend with an exponent close to the value measured across actual brain regions ($\bar{\mu} = 0.84 \pm 0.14$, Fig. 2, bottom right panel shows the collapse for one example region).

Summing up, we find that a very simple linear-rate model tuned to the vicinity of its critical point generates scale-invariant patterns of activity, with associated critical exponents that match remarkably well those observed for the resting-state activity across brain regions in the mouse brain. Understanding whether more elaborate models, including e.g. the Dale's rule in their connectivity matrix, non-linear interactions, spiking neurons, etc. do behave in the same universal way or, rather, represent a better description of the scaling features in actual neural networks in the brain remains as an open task for the future.

4 Unveiling a Link Between the Brain Internal Representations of Stimuli and Criticality

In the previous sections, we have relied on recordings of spontaneous (resting-state) activity to show through different arguments how the dynamics of these networks seems to be poised near a phase transition. However, brains rarely work in such a resting state, but are rather constantly faced with external stimuli that need to be processed.

One needs to be particularly cautious when trying to apply the techniques presented in the previous sections to recordings of input-evoked activity. For instance, the theory that allowed us to estimate the distance to criticality of different brain regions in Sect. 3.3 requires the time-series to be stationary, a property that usually does not hold in input-evoked activity. How much can then be said about the dynamical regime of this type of activity?

To answer this and related questions, we first introduce a method that allows one to separate input-related activity—this is, activity that codifies information about the presented external stimulus—from the "background" activity, which, for practical purposes, we identify here with noise (although, of course, it could emerge from the encoding of other unknown or latent processes).

4.1 A Mathematical Method to Disentangle Input-Related from Background Activity in Neural Recordings

When trying to make sense of high-dimensional data, as in the case of recordings of thousands of neurons, a common and very practical tool to reduce the dimensionality of the problem and find directions of interest is Principal Component Analysis (PCA). Mathematically speaking, given a matrix of observations $X \in \mathbb{R}^{N\times T}$, where N is the number of variables and T the number of samples or observations, PCA seeks to find a transformation or change of basis, P, such that $Y = PX$ is a better representation of the data, meaning that: (i) the redundancy between the variables (i.e., the covariance) is minimized, and (ii) the signal-to-noise ratio is maximized, this is, P projects the original data into the directions of maximal variance. PCA finds a particularly elegant solution: in the new basis, the covariance matrix C_Y must be diagonal, so the projector P is just the matrix of eigenvectors of C_X arranged in columns [67].

Let us now introduce the problem set up by Stringer et al. in [75], where they recorded the response of $N \approx 10{,}000$ neurons in the mouse visual (V1) cortex to the presentation of $T = 2800$ images. If PCA was applied directly over this dataset, one would find the directions of maximal variability, but nothing could be said about the origin of this *overall* variance. In fact, it is estimated that half of the variance of the visual-cortex activity is unrelated to the presented stimuli. Is there a way to tell apart the variance in neural activity steaming from the stimulus encoding ("signal variance"), from the intrinsic or trial-to-trial variability ("noise variance")?

In short, Stringer et al. showed that the stimulus-related variance can indeed be extracted from empirical data by measuring the amount of shared variance between the neural responses to a repeated presentation of the same set of stimuli [75]. The method can then be easily extended to extract not only the signal, but also the noise variance, assuming they span orthogonal sub-spaces within the N-dimensional space [50]. We leave the mathematical details of this method, termed cross-validated PCA (cvPCA), to Appendix 3.

Since cvPCA gives us not only the eigenvalues (input-related variance), but also the associated eigenvectors, it is then possible to project the original activity into an input-related subspace Ψ, independent of the trial, and an orthogonal, trial-dependent subspace Σ_k, such that on trial k:

$$X_k = \Psi + \Sigma_k. \tag{18}$$

Equipped with this method allowing us to disentangle input-related variance of the activity from background fluctuations, let us now see how properties of the former can give us information about the geometry of the internal neural representations.

4.2 Neural Representations Lie in Smooth Manifolds

One way of understanding how the brain encodes information about the external world within the activity of thousands of neurons is through the idea of "neural representations". Given a set of N neurons that we know are involved in the processing of a certain type of stimulus, their average response to one of such inputs can then be understood as a point in the N-dimensional phase space spanning all the possible states of activity.

In a similar way, a set of T similar or related stimuli could be mapped into T points spanning a manifold of dimension $D \leq N$, since one would expect that correlations between the response of the neurons to similar inputs would generate some redundancy in the encoding: i.e., their activities are not expected to be randomly distributed in the whole N-dimensional space, but to occupy a lower-dimensional "representation" manifold [13, 23].

In particular, for a neural encoding to be robust—meaning that representations do not change drastically against small variations of the inputs or perturbations—it is reasonable to assume that these neural manifolds need to be continuous and differentiable (henceforth "smooth"). Remarkably, the authors in [75] presented a mathematical proof that conditioned the "smoothness" of such manifolds to the spectrum of the covariance matrix of neuronal responses to a set of stimulus. More specifically: the eigenvalues of the covariance matrix, when ranked from the largest to the smallest, should decay with their rank according to a power law with an exponent

$$\mu \geq 1 + 2/d, \tag{19}$$

where d represents the embedding dimension of the inputs; if this happens the representation is smooth, and the other way around (details of this proof are not presented here; the interested reader is referred to [75] for an in-depth explanation of this smoothness-to-spectrum relation and its implications). Thus, it is straightforward to see that $\mu = 1$ serves as a lower bound for the eigenspectrum decay-exponent for complex, high-dimensional inputs.

Experimental recordings of over 10,000 individual neurons in the mouse visual cortex exposed to a vast sequence of images with a different intrinsic dimensionality have indeed validated these theoretical predictions [75]. In particular, Stringer et al. showed that the brain is capable of generating internal representations which are optimal in the sense that they are as high-dimensional as possible (i.e., with a slow-decaying covariance spectrum, meaning that there are many principal components to be used for information encoding), while respecting the aforementioned boundary for the smoothness of the representation: $\mu \sim 1 + 2/d$.

4.3 *Optimal Input Representations Arise at the Edge of Instability: A Reservoir-Computing Example*

We have seen that the mouse brain is capable of achieving optimal internal representations of external stimuli, provided these live in smooth manifolds which are as high-dimensional as possible given the constraint presented in the previous section. But, what does this optimality principle, referred to the spectrum of covariance matrices, says about the dynamical regime underlying neural representations? Does it have any relationship or dependency on the underlying critical behavior?

Before trying to answer the above questions, let us remark that there are two caveats preventing us from using the results and methods discussed in the previous sections (Sects. 3.1 and 3.2) for estimating the distance to the edge-of-instability in recordings of input-evoked activity.

1. The aforementioned methods require the timeseries to be stationary, a condition that typically does not hold if the recordings of neural activity are not from resting-state networks, as already mentioned.

2. Thus far, our estimates of the distance to criticality have relied on measures of the degree to which fluctuations of the resting-state activity of neurons are shared (known as spike-count or **noise correlations**), whereas the smoothness condition derived in [75] refers to the spectrum of the **signal correlations**, i.e., the pairwise correlations in the mean responses of neurons to a given set of stimuli.

For these reasons, we need to take a different approach; namely, we will try to recover the phenomenology observed in [75] for neurons in the mouse visual cortex by designing an artificial neural network capable of performing an image-classification task. The interested reader can find detailed explanations of all the forthcoming results in [49].

The following analyses resort on a paradigmatic example of reservoir computing: the Echo State Network (ESN) [31, 39, 77], whose architecture and training for image classification is summarized in Fig. 6. More specifically, the basic structure of ESNs consists of:

- An input layer, which scales a number L_1 of inputs at each time step before they arrive in the reservoir, according to some random weights $W^{in} \in \mathbb{R}^{N \times L_1}$, where N is the number of internal units inside the reservoir.
- A reservoir consisting of N units connected through random weights $W^{res} \in \mathbb{R}^{N \times N}$, whose states evolve according to a non-linear, time-discrete dynamical equation under the influence of a time-dependent input. In this way, the reservoir projects the external input into the high-dimensional space spanned by the activity of its internal units.
- An output layer that converts the information contained in the high-dimensional states of the neurons (i.e., the internal representation of the inputs) to generate the final output.

Unlike other artificial neural networks, the internal weights or "synaptic connections" in ESNs do not need to be updated during the learning process. Training is achieved by just modifying the layer of output weights that readout the network internal states. As we will see in what follows, the internal dynamics of ESNs can be found either in a stable or in a chaotic regime—with a critical point in between—depending on the hyper-parameters of the network (see below).

In order to adapt the ESN architecture—usually employed in time-series analyses—for an image classification task, we used black and white images with $L_1 \times L_2$ pixels (each of them characterized by a value in the [0, 1] interval, representing a normalized gray-scale) and converted them into multivariate time series by considering their vertical dimension as a vector of L_1 elements or features, that "evolve" along $T = L_2$ discrete "time" steps.

One can then define a standard training protocol [5] in which, as illustrated in Fig. 6, at each time $t \in [0, T]$, vectors $\mathbf{u}(t) \in [0, 1]^{L_1}$ corresponding to columns of the image are fed as inputs to the ESN. In this way, the network dynamics for the reservoir states is given by the following non-linear activation function:

$$\mathbf{x}(t) = \tanh(\varepsilon W^{in}\mathbf{u}(t) + W^{res}\mathbf{x}(t-1)) + \xi(t), \qquad (20)$$

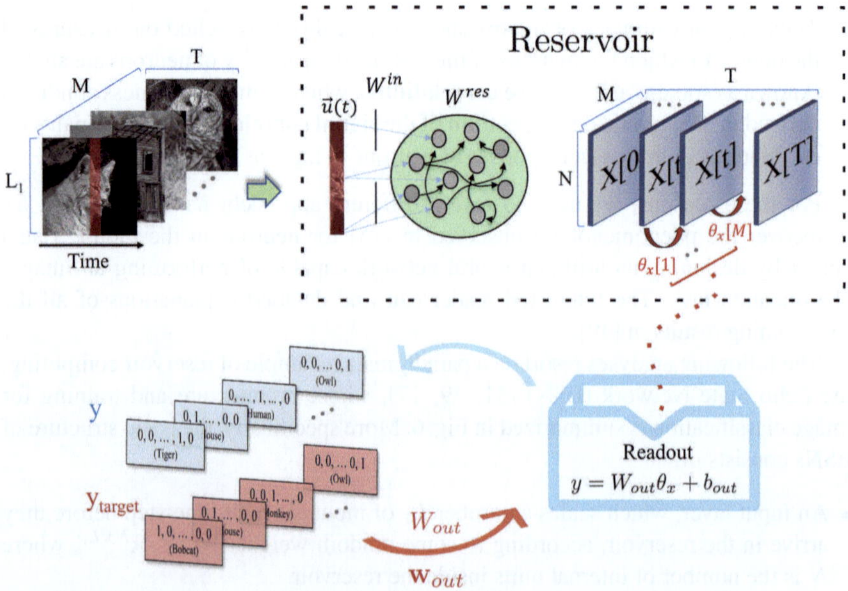

Fig. 6 Training process of an ESN for an image classification task. Images are converted to multivariate time series and then fed into the reservoir. For each processed image a set of parameters θ_x is generated, which characterizes the high-dimensional state of the reservoir, i.e. the "*reservoir model space*". Those parameters are then fed into the readout module, that linearly transforms the information in the reservoir model space into an output label. Finally, output weights W_{out} and biases $\mathbf{b}_{out}$ are generated by minimizing the error between the predicted and target labels. Red arrows indicate steps in which a Ridge regression is performed. Adapted with modifications from [49]

where ε is an input scaling factor and $\xi(t)$ is a white-noise term with zero mean and correlation $\langle \xi_i(t)\xi_j(t')\rangle = \sigma^2\delta_{ij}\delta(t-t')$. Unless otherwise stated, we always consider the deterministic case in which $\sigma = 0$.

Using a supervised learning scheme, the goal of the ESN is to transform (map) the internal representations of the input into a binary output label, $\mathbf{y} \in \mathbb{N}^F$, that correctly classifies each image in the test set as belonging to one of the F existing categories or classes (see Fig. 6):

$$\mathbf{y} = W_{out}\theta_x + \mathbf{b}_{out}, \tag{21}$$

where $W_{out} \in \mathbb{R}^{F\times N(N+1)}$ and $\mathbf{b}_{out} \in \mathbb{R}^F$, defined as the output weights and biases, are determined through a Ridge regression that minimizes the error between the produced and target label for all the presented images in the training set. On the other hand, $\theta_x = [\text{vec}(W_x); \mathbf{b}_x]$ is the "reservoir model space", a set of parameters encoding the reservoir *dynamical* state for a given input (image), obtained from a linear regression to predict the next reservoir state from the past one at discrete time steps:

$$\mathbf{x}(t+1) = W_x\mathbf{x}(t) + \mathbf{b}_x. \tag{22}$$

Thus, once trained, our ESN should be able to decode the identity of the presented input images from the internal representations elicited by such inputs in the reservoir. A direct comparison with the experiment in Stringer et al. naturally raises the following questions:

- Are ESNs able to generate "optimal" internal representations of d-dimensional inputs, which are as high-dimensional as possible while respecting the smoothness boundary given by $\mu > 1 + 2/d$?
- If so, what characterizes the dynamical regime in which these optimal representations emerge?
- Are the dimensionality and smoothness of the internal representations related to the network's ability to classify the input images?

In order to answer the first two questions, we modified two hyperparameters that drastically determine the dynamical regime within the reservoir: the *spectral radius*—largest eigenvalue—of the internal connectivity matrix, W^{res}, which controls the dynamical stability inside the reservoir when no input is fed into the network, and the *scaling factor* ε of the input weights, which can turn an initially expanding mapping into a contracting dynamics, as stronger inputs tend to push the activities of the reservoir units towards the tails of the non-linearity. All the other parameters (number of units, sparsity of connections, weight distribution, etc.) were kept at fixed values (see [49] for further details).

Having set the two knobs that tune the dynamical regime of the ESN, one just needs to define a good dial to figure out how close or far the dynamics is from the edge-of-instability. For this, we used the maximal Lyapunov exponent (MLE), a typical measure of the system sensitivity to perturbations (and, by extension, a way of quantifying chaos), defined as:

$$\text{MLE} = \lim_{k \to \infty} \frac{1}{k} \log \left(\frac{\gamma_k}{\gamma_0} \right) \tag{23}$$

where γ_0 is the initial distance between the perturbed and unperturbed trajectories, and γ_k is the distance between the trajectories at time-step k [7, 72]. Thus, a positive value of the MLE is characteristic of chaotic dynamics, whereas the system is said to be stable to local perturbations provided MLE < 0.

The surface in Fig. 7 (left) shows the exponent of the covariance matrix eigenspectrum as a function of the parameters (ρ, ε), for an ESN presented with the same set of natural images used in [75]. First of all, notice how the boundary exponent for "smooth" high-dimensional input representations, $\mu \approx 1$, coincides with the transition between the stable an unstable regimes (MLE $= 0$, color-coded). This means that the system is able to generate "optimal" representations of the input images, with properties akin to those observed in the visual cortex of mice, provided its dynamics lays close to a critical point!

Now, do these representations translate into a better performance when the system has to carry an input-processing task? Since the original image dataset in Stringer et al. is relatively small and does not have many training examples for the same test

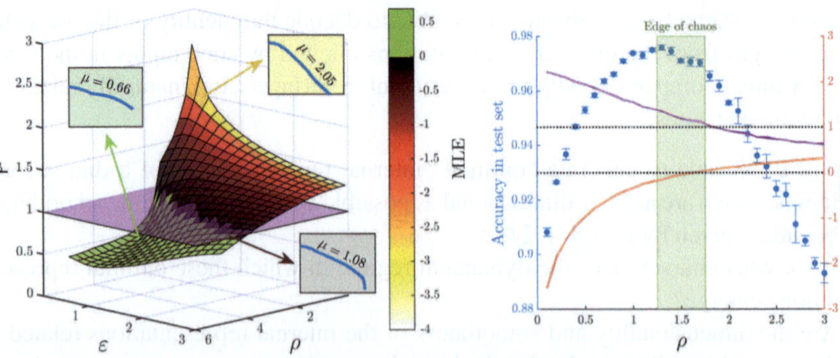

Fig. 7 Optimal performance and smooth, high-dimensional internal representations meet near the edge-of-instability. Left: exponent for the power-law decay of the spectrum of the activity covariance matrix as a function of the spectral radius (ρ) and input scaling factor (ε) of the reservoir, plotted together with the maximum Lyapunov exponent (MLE) color-coded within the surface. The insets correspond to the activity covariance matrix eigenspectrum measured in three different points of the parameter space, where the variance in the n-th dimension (n-th eigenvalue) scales as a power-law $n^{-\mu}$ of the rank. The purple plane marks the boundary $\mu = 1$ for smooth representations of high-dimensional inputs. Right: accuracy in MNIST testset (blue dots), maximum Lyapunov exponent (orange line) and best fit exponent for the power-law spectrum of the activity covariance matrix (purple line). Errors were estimated as the standard deviation over ten different realizations of the ESN. Adapted with modifications from [49]

labels, to answer this question we relied on a classification task over the MNIST dataset, a common benchmarking dataset consisting in hand-written digits. In Fig. 7 (right) we analyzed the percentage of images that were correctly classified during the test as a function of the spectral radius, while also recording the MLE as a measure of distance to the edge-of-instability. Remarkably, optimal performance for this classification task is obtained when the system is approaching the edge-of-instability from the stable regime, concomitant with a exponent $\mu \approx 1$ for the eigenspectrum of the activity covariance matrix.

To conclude, it is important to remark that the inputs fed to the network to obtain the previous results were natural images and MNIST digits, and thus can be considered high-dimensional (i.e., $d \to \infty$), so that the upper bound in the covariance spectrum exponent for the smoothness of the manifold is given by $\mu \geq 1 + 2/d \approx 1$. Should we expect to find a similar match between edge-of-instability dynamics and optimal, smooth representations of low-dimensional inputs?

Following the experimental protocol of Stringer et al., we constructed 8-dimensional and 4-dimensional inputs by projecting the natural images used in [75] into their first $d = 8$ and $d = 4$ principal component, respectively. Figure 8 shows the rank-ordered covariance eigenvalues of the reservoir states, together with the best power-law fit exponents, for natural and low-dimensional inputs. For the sake of comparison, we also reproduced in yellow the eigenvalue rank-ordered distributions obtained from the experimental data on mouse visual cortex [75]. Not only our simulations are in perfect agreement with the experimental results, but one can

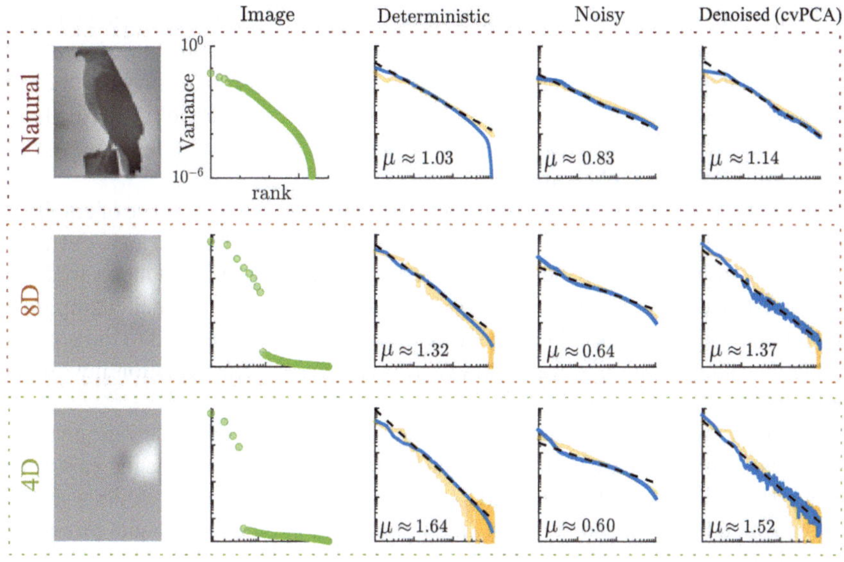

Fig. 8 ESNs near the edge-of-instability generate representations with the same eigenvalue distributions observed in real neurons. From left to right: one sample from the $M = 2800$ images in the training set; covariance eigenspectrum of the images pixel intensities; covariance eigenspectrum of the internal states of an ESN (blue line) and real, V1 mouse neurons (yellow line, plotted after [75] applying cvPCA) when subject to images of dimensionality d ; same analysis, but now zero-mean white noise of amplitude $\epsilon = 0.4$ is added to the neuron dynamics (blue line), and no cvPCA is performed over the experimental recordings (yellow line); same analysis as in the previous panel, but now noise has been subtracted using cvPCA in both, simulations and experiments. From top to bottom: results for natural, high-dimensional images; the same images projected onto 8 dimensions; the same images projected onto 4 dimensions. To obtain the ESNs eigenspectra, parameters were chosen so that the networks operated near the edge-of-instability, with MLE $\sim -5 \cdot 10^{-3}$. Adapted with modifications from [49]

also add a small noise $\xi \neq 0$ into the internal dynamics of the reservoir units, recovering the observed exponents before isolating the input-related variance through a cross-validated PCA (cvPCA) analysis, as explained in Sect. 4.1.

Therefore, summing up, the artificial neural network that we have designed under the paradigm of reservoir computing is able to construct optimal representation of the inputs when it operates close (but slightly below) to criticality (the edge of instability). It is precisely under such conditions the network performance is optimal.

5 Summary, Discussions, and Open Directions

We have presented a large diversity of analyses of neural networks, both biological and artificial, trying to shed light on the relationships between their internal dynamical regimes and their emergent or collective computational capabilities.

First of all, we observed the scale-invariant nature of neural activity in the mouse brain emerge under a phenomenological renormalization group flow, resembling closely what happens at critical points of second-order phase transitions. Moreover, the so-determined scaling exponents exhibit little (yet some) variability across brain regions, so we referred to them as been "quasi-universal", in seeming analogy with universality in critical phenomena.

After these scaling analyses, we set out to quantify the actual distance of empirical data to the edge of instability, finding that all 16 regions analyzed turn out to be relatively close to such a critical point. Interpreting the reason behind the variability in the distance to criticality across regions remains an open and challenging goal. In particular, we have speculated that this variability could be related to the position of the considered region in the hierarchical organization of information flow in the brain [50].

To close the loop we constructed a very simple model of linear-rate neurons, which can exhibit a phase transition as the overall coupling strength is increased. We found that this model, when tuned to the vicinity of its critical point, exhibits scaling as elucidated by the phenomenological renormalization group technique. Moreover, the scaling exponents are surprisingly close to those found in actual brain activity, which also points out in the direction of possible universality. Analyzing whether this behavior is really universal in the strict sense of statistical physics or if it is approximate remains as an open challenge to be clarified in future work.

In summary, we report on solid quantitative evidence that actual neural activity in the mouse brain seems to operate, rather generically, close to a critical point or point of instability. Thus, the next question to be posed is: what is the reason for that?

Even if the potential functional advantages of critical behavior in both real and artificial neural networks [7, 29, 66] (as well as, in general, on biological systems, [48, 52]), have been extensively explored in recent years, here we go a step further and aimed at finding a relation between critical behavior and optimal representation of external inputs, i.e., optimal computational capabilities.

For this, we first recapitulated a recent seminal contribution by Stringer et al. [75]. They first devised a method to separate variance stemming from inputs (signal covariance) from intrinsic fluctuations (noise covariance). Then, they derived a rather elegant mathematical theorem establishing that optimal representations of inputs within a network of neurons should be constructed on lower-dimensional manifolds in the space of neural activity, in which the corresponding spectrum of the (signal) covariance matrix should decay with an exponent (μ) slightly larger that 1, which is larger for lower-dimensional (simpler) inputs (i.e. $\mu \approx 1 + 2/d$, where d is the input dimensionality).

All this might seem very abstract, however, it provides us with a quantitative tool to scrutinize how close to optimality are actual patterns of neural activity while encoding inputs. Indeed, by analyzing high-quality data from the mouse visual cortex while the mouse was exposed to a series of visual stimuli, Stringer et al. achieved the very remarkable conclusion that, as a matter of fact, empirical data describe neural representations that are very close to the predicted optimal behavior, i.e., the spectrum of the signal covariance matrix decays with an exponent slightly larger than one, that increases as the set of stimuli becomes simpler (lower dimensional).

Therefore, on the one hand, we have found evidence of scaling and close to critical behavior across brain areas and, on the other hand, we have found that the visual cortex seems to operate in a regime allowing it to generate optimal representations of stimuli. Thus, we wondered whether these two features come generically hand by hand.

To shed light onto this last question we constructed echo-state networks, which are one of the best known examples of reservoir computing. This is a paradigm of computation different from the standard von-Neumann's one, allowing to exploit the collective dynamical state of the neural network (the reservoir) in a generic and flexible way, without modifying the synaptic weights between the computational units of the network during training.

The internal state of the network can then be tuned by changing some global parameter (such as, e.g., the overall synaptic strength) and there is a critical point where the dynamics becomes unstable. We found that, right at such a critical point, the network exposed to series of inputs (actually, the very same images used by Stringer et al. in [75]), produces optimal representations (i.e. $\mu \approx 1$), which are as high-dimensional as possible while remaining continuous and differentiable. Moreover, and rather remarkably, when trained to perform a classification task on the digits from the MNIST dataset, the performance of the network is found to be optimal when its dynamics is slightly below this edge of instability.

The corollary of all this is that neural networks—both biological and artificial—seem to behave in an optimal fashion in what respects input representation and computational performance if they are tuned to be slightly below the edge of instability, i.e. when they are close to criticality. This result adds to the mounting evidence revealing that criticality is a key concept for the understanding of neural networks as well as other information-processing biological systems [52].

As a matter of fact, in recent years, there have been some attempts to design general-purpose neuromorphic computation devices (chips) exploiting the potential advantages of criticality (see, e.g., [16, 32, 43, 74]), a direction that we find most suggestive and promising.

We hope that more synergies between neuroscience and machine learning emerge from these analyses. These should help us deciphering the meaning and functionality of brain activity as well as designing novel and powerful artificial-intelligence paradigms inspired more closely by biological networks.

Acknowledgements We acknowledge the Spanish Ministry and Agencia Estatal de investigación (AEI) through Project of I+D+i Ref. PID2020-113681GB-I00, financed by MICIN/AEI/10.13039/501100011033 and FEDER "A way to make Europe", as well as the Consejería de Conocimiento, Investigación Universidad, Junta de Andalucía and European Regional Development Fund, Project reference P20-00173 for financial support. We also thank R. Calvo, C. Martorell, V. Buendía, P. Villegas, R. Corral, J. Pretel, P. Moretti, and M. Ibañez, for valuable discussions and/or suggestions.

Appendix 1: Relating Covariances to Connectivity

To prove Eq. (13) for the linear-rate model, one just needs to take the trial-averaged (average over realizations of $\xi_i(t)$) firing rate at the stationary state. In the frequency domain and particularizing already for the case in which $\omega = 0$ (for which C_{ij} can be easily interpreted as the long-time window or spike-count covariance):

$$\langle x_i(\omega = 0)\rangle = (I - J)^{-1}\langle \xi_i(\omega = 0)\rangle, \tag{24}$$

where $\langle\cdot\rangle$ denotes an average over trials. Then the matrix $C(0)$, i.e. the matrix of the Fourier transform of the covariances at zero frequency, is given by:

$$C_{ij}(0) = \langle x_i(0)x_j(0)\rangle = (I - J)^{-1}((I - J)^{-1})^T. \tag{25}$$

This relation—that has been proven to be remarkably general—holds as a very good approximation for many different (nonlinear) models, including spiking neural networks (see [17, 54, 58, 81] for a recent review).

Appendix 2: Rank-Ordered Eigenvalues Versus Probability Densities

In Sect. 2.1 we argued that when correlations are self-similar inside a cluster of size K, the n-th largest eigenvalue λ_n decays with its rank n (when ordered from the largest to the smallest) as:

$$\lambda_n = A\left(\frac{n}{K}\right)^{-\mu}. \tag{26}$$

On the other hand, in Sect. 3.2 we reported on the probability density of the eigenvalues of the covariance matrix, Eq. (15), and argued that its tail converges to a power-law when the system is close to the edge of instability:

$$p(\lambda) \sim B\lambda^{-\nu}. \tag{27}$$

Here we make explicit the relationship between these two power-laws.

The spectrum of a matrix of size K is composed by K discrete eigenvalues:

$$p(\lambda) = \frac{1}{K}\sum_{n=1}^{K}\delta(\lambda - \lambda_n). \tag{28}$$

In the limit of $K \to \infty$ the above discrete distribution can be approximated by a continuous one:

$$p_{RC}^{(c)}(\lambda) \approx \frac{1}{K}\int_1^{\infty} \delta\left(\lambda - \lambda_n\right) dn = \frac{1}{K}\int_1^{\infty} \delta\left(\lambda - A\left(\frac{n}{K}\right)^{-\mu}\right) dn,$$

where we used Eq. (26) for the rank-ordering of the spectrum . The function $g(n) \equiv \lambda - A\left(\frac{n}{K}\right)^{-\mu}$ has roots $n_0 = K\left(\frac{\lambda}{A}\right)^{-\frac{1}{\mu}}$. Using the following property of the Dirac-delta distribution:

$$\delta\left(g(n)\right) = \frac{\delta\left(n - n_0\right)}{\left|g'\left(n_0\right)\right|} = \frac{\delta\left(n - n_0\right)}{K^{-1}A\mu\left(\frac{\lambda}{A}\right)^{\frac{\mu+1}{\mu}}}, \tag{29}$$

the spectral density reads:

$$p_{RC}^{(c)}(\lambda) \approx \frac{1}{K\mu} A^{\frac{1}{\mu}} K \lambda^{-1-\frac{1}{\mu}} \int_1^{\infty} \delta\left(n - n_0\right) dn = \frac{1}{\mu} A^{\frac{1}{\mu}} \lambda^{-1-\frac{1}{\mu}}. \tag{30}$$

Thus, by comparing Eq. (27) with Eq. (30), one readily finds a relationship between the power law exponent μ of the rank-ordered eigenvalues and the exponent of the spectral distribution ν:

$$\nu = 1 + \frac{1}{\mu}. \tag{31}$$

The reader can find a more detailed discussion of this scaling relation in, e.g., [18, 35].

Appendix 3: Cross-validate PCA

Following Stringer et al. [75], let us consider two observation matrices, $X_{(1)}, X_{(2)} \in \mathbb{R}^{N\times T}$, corresponding to two identical realizations or trials of an experiment. For simplicity, we further assume that the mean activity of each neuron across images has been subtracted, so that both matrices have zero-mean rows. The Singular Value Decomposition (SVD) theorem states that any observation matrix X can be decomposed as:

$$X = USV^T, \tag{32}$$

so that

$$XV = US \Longrightarrow X\mathbf{v}_i = \sigma_i \mathbf{u}_i, \tag{33}$$

where $U \in \mathbb{R}^{N\times r}$ contains the $\mathbf{u}_i$ eigenvectors by columns, $V \in \mathbb{R}^{T\times r}$ contains the $\mathbf{v}_i$ eigenvectors by columns, and $S \in \mathbb{R}^{r\times r}$ is a diagonal matrix containing the singular values $s_{ii} = \sigma_i = \sqrt{\lambda_i}$.

Assume one has performed SVD on the first trial observation matrix $X_{(1)}$, obtaining the singular vectors $U_{(1)}$ and $V_{(1)}$. Defining the projection matrix $P = U_{(1)}^T \in \mathbb{R}^{r\times N}$, which diagonalizes $C_u^{(1)} = \frac{1}{T-1} X_{(1)} X_{(1)}^T$, one can compute:

$$Y_{(1)} = PX_{(1)} \Longrightarrow \mathbf{y}_i^{(1)} = (\mathbf{u}_i^{(1)})^T X_{(1)} \in \mathbb{R}^{1\times T}, \tag{34}$$

where $Y_{(1)} \in \mathbb{R}^{r\times T}$ contains the projection of the first trial activity over its r principal components. Now, the idea of cv-PCA is to project on the same subspace the observations coming from the second trial [75]:

$$Y_{(2)} = PX_{(2)} \Longrightarrow \mathbf{y}_i^{(2)} = (\mathbf{u}_i^{(1)})^T X_{(2)} \in \mathbb{R}^{1\times T}, \tag{35}$$

and ask to form a covariance matrix with the product of both projections:

$$C_\psi = \frac{1}{T-1} Y_{(1)} Y_{(2)}^T = \frac{1}{T-1} \begin{pmatrix} \mathbf{y}_1^{(1)} \cdot \mathbf{y}_1^{(2)} & \mathbf{y}_1^{(1)} \cdot \mathbf{y}_2^{(2)} & \cdots \\ \mathbf{y}_2^{(1)} \cdot \mathbf{y}_1^{(2)} & \mathbf{y}_2^{(1)} \cdot \mathbf{y}_2^{(2)} & \cdots \\ \cdots & \cdots & \mathbf{y}_r^{(1)} \cdot \mathbf{y}_r^{(2)} \end{pmatrix}. \tag{36}$$

One can now hypothesize that the activity at time t of any neuron i during trial k can be linearly decomposed as:

$$\mathbf{x}_i^{(k)}(t) = \psi_i(t) + \epsilon_i^{(k)}(t), \tag{37}$$

where ψ_i denotes the input-related activity, which should be independent of the trial for experiments carried in identical conditions; and $\epsilon_i^{(k)}$ is defined as background or trial-to-trial variable activity, spanning a subspace that is orthogonal to the input-related one. Let us now show that the i-th diagonal element of C_ψ is a non-biased estimator of the input-related variance ω_i:

$$\omega_i = \frac{1}{T-1} \mathbf{y}_i^{(1)} \cdot \mathbf{y}_i^{(2)} = \frac{1}{T-1} (\mathbf{u}_i^{(1)})^T X_{(1)} X_{(2)}^T \mathbf{u}_i^{(1)}. \tag{38}$$

Rewriting the observation matrices as $X_{(k)} = \Psi + \Sigma_{(k)}$, where Ψ contains vectors ψ_i in rows and so does $\Sigma_{(k)}$ with vectors $\epsilon_i^{(k)}$:

$$\omega_i = \frac{1}{T-1} \mathbf{u}_i^T \left(\Psi + \Sigma_{(1)}\right) \left(\Psi + \Sigma_{(2)}\right)^T \mathbf{u}_i = \frac{1}{T-1} \mathbf{u}_i^T \Psi \Psi^T \mathbf{u}_i + \frac{1}{T-1} \mathbf{u}_i^T \Sigma_{(1)} \Psi^T \mathbf{u}_i, \tag{39}$$

where all the terms containing $\Sigma_{(2)}$ can be dismissed due to the statistical independence of the realizations. Under reasonable approximations shown in [75], one can prove that if the singular vectors $\mathbf{u}_i$ approach the singular vectors of Ψ, then the first term converges to the actual variance of the input-related activity along the principal direction $\mathbf{u}_i$, while the second term converges to zero. We have thus seen how the *cross-validated PCA* (cvPCA) method proposed by Stringer et al. allows one to estimate the input-related covariances.

Is it possible to take a step further and find a proxy for the input-related activity observation matrix $\Psi \in \mathbb{R}^{N\times T}$, such that $\tilde{C}_\psi = \frac{1}{T-1} \Psi \Psi^T$ has the same eigenvalue spectrum as C_ψ? To do so, we first define a new basis of projected vectors given by:

$$\mathbf{z}_i = \sqrt{\mathbf{y}_i^{(1)} \cdot \mathbf{y}_i^{(2)}} \frac{\mathbf{y}_i^{(2)}}{\left\| \mathbf{y}_i^{(2)} \right\|} \in \mathbb{R}^{1\times T}. \tag{40}$$

which clearly fulfills $\mathbb{E}\left[\mathbf{z}_i \cdot \mathbf{z}_i^T\right] = \omega_i$ under the same conditions as above. If $Z \in \mathbb{R}^{r\times T}$ is the matrix composed by the r vectors $\mathbf{z}_i$ in rows, then:

$$\Psi = P^T Z \tag{41}$$

is the input-related activity of the neurons. From there, it is straightforward to see, using Eq. (37), that one can estimate the background activity just by subtracting the input-related activity from the raw data:

$$\Sigma_{(k)} = X_{(k)} - \Psi. \tag{42}$$

References

1. Arieli, A., Sterkin, A., Grinvald, A., Aertsen, A.: Dynamics of ongoing activity: explanation of the large variability in evoked cortical responses. Science **273**(5283), 1868 (1996)
2. Beggs, J.M., Plenz, D.: Neuronal avalanches in neocortical circuits. J. Neurosci. **23**(35), 11167–11177 (2003)
3. Beggs, J.M., Timme, N.: Being critical of criticality in the brain. Front. Physiol. **3**, 163 (2012)
4. Bertschinger, N., Natschläger, T.: Real-time computation at the edge of chaos in recurrent neural networks. Neural Comput. **16**(7), 1413–1436 (2004)
5. Bianchi, F.M., Scardapane, S., Løkse, S., Jenssen, R.: Reservoir computing approaches for representation and classification of multivariate time series. IEEE Trans. Neural Netw. Learn. Syst. (Conference) **32**(5), 2169–2179 (2021). https://doi.org/10.1109/TNNLS.2020.3001377
6. Binney, J.J., Dowrick, N.J., Fisher, A.J., Newman, M.E.: The Theory of Critical Phenomena: An Introduction to the Renormalization Group. Oxford University Press (1992)
7. Boedecker, J., Obst, O., Lizier, J.T., Mayer, N.M., Asada, M.: Information processing in echo state networks at the edge of chaos. Theory Biosci. **131**(3) (2011). https://doi.org/10.1007/s12064-011-0146-8
8. Boedecker, J., Obst, O., Lizier, J.T., Mayer, N.M., Asada, M.: Information processing in echo state networks at the edge of chaos. Theory Biosci. **131**(3), 205–213 (2012)
9. Brunel, N.: Dynamics of sparsely connected networks of excitatory and inhibitory spiking neurons. J. Comput. Neurosci. **8**(3), 183–208 (2000)
10. Buendía, V., Villegas, P., Di Santo, S., Vezzani, A., Burioni, R., Muñoz, M.A.: Jensen's force and the statistical mechanics of cortical asynchronous states. Sci. Rep. **9**(1), 15183 (2019)
11. Cabral, J., Kringelbach, M.L., Deco, G.: Functional connectivity dynamically evolves on multiple time-scales over a static structural connectome: models and mechanisms. NeuroImage **160**, 84–96 (2017)
12. Chialvo, D.R.: Emergent complex neural dynamics. Nat. Phys. **6**(10), 744–750 (2010)
13. Chung, S., Abbott, L.: Neural population geometry: an approach for understanding biological and artificial neural networks. Curr. Opin. Neurobiol. **70**, 137–144 (2021)
14. Cocchi, L., Gollo, L.L., Zalesky, A., Breakspear, M.: Criticality in the brain: a synthesis of neurobiology, models and cognition. Progr. Neurobiol. (2017)
15. Cocco, S., Monasson, R., Zamponi, F.: From Statistical Physics to Data-driven Modelling: With Applications to Quantitative Biology. Oxford University Press (2022)

16. Cramer, B., Stöckel, D., Kreft, M., Wibral, M., Schemmel, J., Meier, K., Priesemann, V.: Control of criticality and computation in spiking neuromorphic networks with plasticity. Nat. Commun. **11**(1), 2853 (2020)
17. Dahmen, D., Grün, S., Diesmann, M., Helias, M.: Second type of criticality in the brain uncovers rich multiple-neuron dynamics. Proc. Natl. Acad. Sci. **116**(26), 13051–13060 (2019)
18. De Marzo, G., Gabrielli, A., Zaccaria, A., Pietronero, L.: Dynamical approach to Zipf's law. Phys. Rev. Res. **3**(1), 013084 (2021)
19. Deco, G., Jirsa, V.K.: Ongoing cortical activity at rest: criticality, multistability, and ghost attractors. J. Neurosci. **32**(10), 3366–3375 (2012)
20. Deco, G., Jirsa, V.K., Robinson, P.A., Breakspear, M., Friston, K.: The dynamic brain: from spiking neurons to neural masses and cortical fields. PLoS Comput. Biol. **4**(8), e1000092 (2008)
21. Demas, J., Manley, J., Tejera, F., Barber, K., Kim, H., Traub, F.M., Chen, B., Vaziri, A.: High-speed, cortex-wide volumetric recording of neuroactivity at cellular resolution using light beads microscopy. Nat. Methods **18**(9), 1103–1111 (2021)
22. Efrati, E., Wang, Z., Kolan, A., Kadanoff, L.P.: Real-space renormalization in statistical mechanics. Rev. Mod. Phys. **86**(2), 647 (2014)
23. Ganguli, S., Sompolinsky, H.: Compressed sensing, sparsity, and dimensionality in neuronal information processing and data analysis. Annu. Rev. Neurosci. **35**, 485–508 (2012)
24. Ginibre, J.: Statistical ensembles of complex, quaternion, and real matrices. J. Math. Phys. **6**(3), 440–449 (1965)
25. Girko, V.L.: The Circular Law. Twenty Years Later. Part III (2005)
26. Hage, T.A., Bosma-Moody, A., Baker, C.A., Kratz, M.B., Campagnola, L., Jarsky, T., Zeng, H., Murphy, G.J.: Synaptic connectivity to l2/3 of primary visual cortex measured by two-photon optogenetic stimulation. Elife **11**, e71103 (2022)
27. Hahn, G., Petermann, T., Havenith, M.N., Yu, S., Singer, W., Plenz, D., Nikolić, D.: Neuronal avalanches in spontaneous activity in vivo. J. Neurophysiol. **104**(6), 3312–3322 (2010)
28. Haimovici, A., Tagliazucchi, E., Balenzuela, P., Chialvo, D.R.: Brain organization into resting state networks emerges at criticality on a model of the human connectome. Phys. Rev. Lett. **110**, 178101 (2013)
29. Hidalgo, J., Grilli, J., Suweis, S., Munoz, M.A., Banavar, J.R., Maritan, A.: Information-based fitness and the emergence of criticality in living systems. Proc. Natl. Acad. Sci. **111**(28), 10095–10100 (2014)
30. Hu, Y., Sompolinsky, H.: The spectrum of covariance matrices of randomly connected recurrent neuronal networks with linear dynamics. PLoS Comput. Biol. **18**(7), e1010327 (2022)
31. Jaeger, H.: The "echo stateâŁž approach to analysing and training recurrent neural networks-with an erratum note. Bonn, Germany: German National Research Center for Information Technology GMD Technical Report **148**(34), 13 (2001)
32. Kuncic, Z., Nakayama, T.: Neuromorphic nanowire networks: principles, progress and future prospects for neuro-inspired information processing. Adv. Phys. X **6**(1), 1894234 (2021)
33. Langton, C.G.: Computation at the edge of chaos: phase transitions and emergent computation. Phys. D Nonlinear Phenom. **42**(1–3), 12–37 (1990)
34. Li, J., Shew, W.L.: Tuning network dynamics from criticality to an asynchronous state. PLoS Comput. Biol. **16**(9), e1008268 (2020)
35. Li, W.: Zipf's law everywhere. Glottometrics **5**(2002), 14–21 (2002)
36. Liang, J., Zhou, T., Zhou, C.: Hopf bifurcation in mean field explains critical avalanches in excitation-inhibition balanced neuronal networks: a mechanism for multiscale variability. Front. Syst. Neurosci. **14**, 87 (2020)
37. López, R.C., Buendía, V., Muñoz, M.A.: Excitatory-inhibitory branching process: a parsimonious view of cortical asynchronous states, excitability, and criticality. Phys. Rev. Res. **4**(4), L042027 (2022)
38. Lukoševičius, M., Jaeger, H.: Reservoir computing approaches to recurrent neural network training. Comput. Sci. Rev. **3**(3), 127–149 (2009)
39. Lukoševičius, M.: A practical guide to applying echo state networks. In: Montavon, G., Orr, G.B., Müller, K.R. (eds.) Neural Networks: Tricks of the Trade. Lecture Notes in Computer Science, 2nd edn., pp. 659–686. Springer, Berlin, Heidelberg (2012)

40. Ma, Z., Turrigiano, G.G., Wessel, R., Hengen, K.B.: Cortical circuit dynamics are homeostatically tuned to criticality in vivo. Neuron **104**(4), 655–664 (2019)
41. Maass, W., Natschläger, T., Markram, H.: Real-time computing without stable states: a new framework for neural computation based on perturbations. Neural Comput. **14**(11), 2531–2560 (2002)
42. Magnasco, M.O., Piro, O., Cecchi, G.A.: Self-tuned critical anti-Hebbian networks. Phys. Rev. Lett. **102**(25), 258102 (2009)
43. Mallinson, J., Shirai, S., Acharya, S., Bose, S., Galli, E., Brown, S.: Avalanches and criticality in self-organized nanoscale networks. Sci. Adv. **5**(11), eaaw8438 (2019)
44. Martinello, M., Hidalgo, J., Maritan, A., Di Santo, S., Plenz, D., Muñoz, M.A.: Neutral theory and scale-free neural dynamics. Phys. Rev. X **7**(4), 041071 (2017)
45. Massobrio, P., de Arcangelis, L., Pasquale, V., Jensen, H.J., Plenz, D.: Criticality as a signature of healthy neural systems. Front. Syst. Neurosci. **9**, 22 (2015)
46. Mehta, M.: Random matrices and the statistical theory of energy levels. Acad. Press, New York (1967)
47. Meshulam, L., Gauthier, J.L., Brody, C.D., Tank, D.W., Bialek, W.: Coarse graining, fixed points, and scaling in a large population of neurons. Phys. Rev. Lett. **123**(17), 178103 (2019)
48. Mora, T., Bialek, W.: Are biological systems poised at criticality? J. Stat. Phys. **144**, 268–302 (2011)
49. Morales, G.B., Muñoz, M.A.: Optimal input representation in neural systems at the edge of chaos. Biology **10**(8), 702 (2021)
50. Morales, G.B., di Santo, S., Muñoz, M.A.: Quasiuniversal scaling in mouse-brain neuronal activity stems from edge-of-instability critical dynamics. Proc. Natl. Acad. Sci. **120**(9), e2208998120 (2023)
51. Morrell, M.C., Sederberg, A.J., Nemenman, I.: Latent dynamical variables produce signatures of spatiotemporal criticality in large biological systems. Phys. Rev. Lett. **126**(11), 118302 (2021)
52. Muñoz, M.A.: Colloquium: criticality and dynamical scaling in living systems. Rev. Mod. Phys. **90**(3), 031001 (2018)
53. Nicoletti, G., Suweis, S., Maritan, A.: Scaling and criticality in a phenomenological renormalization group. Phys. Rev. Res. **2**(2), 023144 (2020). Publisher: American Physical Society
54. Ocker, G.K., Hu, Y., Buice, M.A., Doiron, B., Josić, K., Rosenbaum, R., Shea-Brown, E.: From the statistics of connectivity to the statistics of spike times in neuronal networks. Curr. Opin. Neurobiol. **46**, 109–119 (2017)
55. O'Byrne, J., Jerbi, K.: How critical is brain criticality? Trends Neurosci. (2022)
56. Packard, N.H.: Adaptation Toward the Edge of Chaos. Center for Complex Systems Research, University of Illinois, Urbana-Champaign (1988)
57. Palva, J.M., Zhigalov, A., Hirvonen, J., Korhonen, O., Linkenkaer-Hansen, K., Palva, S.: Neuronal long-range temporal correlations and avalanche dynamics are correlated with behavioral scaling laws. Proc. Natl. Acad. Sci. **110**(9), 3585–3590 (2013)
58. Pernice, V., Staude, B., Cardanobile, S., Rotter, S.: How structure determines correlations in neuronal networks. PLOS Comput. Biol. **7**(5), e1002059 (2011). Publisher: Public Library of Science
59. Plenz, D., Niebur, E.: Criticality in Neural Systems. Wiley, New York (2014)
60. Plenz, D., Ribeiro, T.L., Miller, S.R., Kells, P.A., Vakili, A., Capek, E.L.: Self-organized criticality in the brain. Front. Phys. **9**, 639389 (2021)
61. Poil, S.S., Hardstone, R., Mansvelder, H.D., Linkenkaer-Hansen, K.: Critical-state dynamics of avalanches and oscillations jointly emerge from balanced excitation/inhibition in neuronal networks. J. Neurosci. **32**(29), 9817–9823 (2012)
62. Ponce-Alvarez, A., Jouary, A., Privat, M., Deco, G., Sumbre, G.: Whole-brain neuronal activity displays crackling noise dynamics. Neuron **100**(6), 1446–1459 (2018)
63. Raichle, M.E.: The restless brain. Brain Connect. **1**(1), 3–12 (2011)
64. Rajan, K., Abbott, L.F.: Eigenvalue spectra of random matrices for neural networks. Phys. Rev. Lett. **97**, 188104 (2006)

65. Renart, A., De La Rocha, J., Bartho, P., Hollender, L., Parga, N., Reyes, A., Harris, K.D.: The asynchronous state in cortical circuits. Science **327**(5965), 587–590 (2010)
66. Shew, W.L., Plenz, D.: The functional benefits of criticality in the cortex. The Neuroscientist **19**(1), 88–100 (2013)
67. Shlens, J.: A Tutorial on Principal Component Analysis. arXiv:1404.1100 [cs, stat] (2014)
68. Softky, W.R., Koch, C.: The highly irregular firing of cortical cells is inconsistent with temporal integration of random EPSPs. J. Neurosci. **13**(1), 334–350 (1993)
69. Solovey, G., Alonso, L.M., Yanagawa, T., Fujii, N., Magnasco, M.O., Cecchi, G.A., Proekt, A.: Loss of consciousness is associated with stabilization of cortical activity. J. Neurosci. **35**(30), 10866–10877 (2015)
70. Sompolinsky, H., Crisanti, A., Sommers, H.J.: Chaos in random neural networks. Phys. Rev. Lett. **61**(3), 259 (1988)
71. Sornette, D.: Critical Phenomena in Natural Sciences: Chaos, Fractals, Selforganization and Disorder: Concepts and Tools. Springer Science & Business Media (2006)
72. Sprott, J.C.: Chaos and Time-series Analysis. Oxford University Press (2003). Google-Books-ID: SEDjdjPZ158C
73. Steinmetz, N.A., Aydin, C., Lebedeva, A., Okun, M., Pachitariu, M., Bauza, M., Beau, M., Bhagat, J., Böhm, C., Broux, M., et al.: Neuropixels 2.0: a miniaturized high-density probe for stable, long-term brain recordings. Science **372**(6539), eabf4588 (2021)
74. Stieg, A.Z., Avizienis, A.V., Sillin, H.O., Martin-Olmos, C., Aono, M., Gimzewski, J.K.: Emergent criticality in complex turing b-type atomic switch networks (2012)
75. Stringer, C., Pachitariu, M., Steinmetz, N., Carandini, M., Harris, K.D.: High-dimensional geometry of population responses in visual cortex. Nature **571**(7765), 361–365 (2019)
76. Tagliazucchi, E., Balenzuela, P., Fraiman, D., Chialvo, D.R.: Criticality in large-scale brain fMRI dynamics unveiled by a novel point process analysis. Front. Physiol. **3**, 15 (2012)
77. Tanaka, G., Yamane, T., Héroux, J.B., Nakane, R., Kanazawa, N., Takeda, S., Numata, H., Nakano, D., Hirose, A.: Recent advances in physical reservoir computing: a review. Neural Netw. **115**, 100–123 (2019)
78. Tao, T., Vu, V., Krishnapur, M.: Random matrices: universality of ESDs and the circular law (2010)
79. Tkačik, G., Mora, T., Marre, O., Amodei, D., Palmer, S.E., Berry, M.J., Bialek, W.: Thermodynamics and signatures of criticality in a network of neurons. Proc. Natl. Acad. Sci. **112**(37), 11508–11513 (2015)
80. Touboul, J., Destexhe, A.: Power-law statistics and universal scaling in the absence of criticality. Phys. Rev. E **95**(1), 012413 (2017)
81. Trousdale, J., Hu, Y., Shea-Brown, E., Josić, K.: Impact of network structure and cellular response on spike time correlations. PLOS Comput. Biol. **8**(3), e1002408 (2012)
82. Van Vreeswijk, C., Sompolinsky, H.: Chaos in neuronal networks with balanced excitatory and inhibitory activity. Science **274**(5293), 1724–1726 (1996)
83. Villegas, P., Moretti, P., Muñoz, M.A.: Frustrated hierarchical synchronization and emergent complexity in the human connectome network. Sci. Rep. **4**(1), 1–7 (2014)
84. Wang, R., Lin, P., Liu, M., Wu, Y., Zhou, T., Zhou, C.: Hierarchical connectome modes and critical state jointly maximize human brain functional diversity. Phys. Rev. Lett. **123**(3), 038301 (2019)
85. Wilson, K.G.: The renormalization group and critical phenomena. Rev. Mod. Phys. **55**(3), 583 (1983)
86. Wilting, J., Priesemann, V.: 25 years of criticality in neuroscience-established results, open controversies, novel concepts. Curr. Opin. Neurobiol. **58**, 105–111 (2019)
87. Winding, M., Pedigo, B.D., Barnes, C.L., Patsolic, H.G., Park, Y., Kazimiers, T., Fushiki, A., Andrade, I.V., Khandelwal, A., Valdes-Aleman, J., et al.: The connectome of an insect brain. Science **379**(6636), eadd9330 (2023)

88. Yang, H., Shew, W.L., Roy, R., Plenz, D.: Maximal variability of phase synchrony in cortical networks with neuronal avalanches. J. Neurosci. **32**(3), 1061–1072 (2012)
89. Zong, W., Obenhaus, H.A., Skytøen, E.R., Eneqvist, H., de Jong, N.L., Vale, R., Jorge, M.R., Moser, M.B., Moser, E.I.: Large-scale two-photon calcium imaging in freely moving mice. Cell **185**(7), 1240–1256 (2022)

Emergent Complex Dynamics in Neuronal Cultures and Its Relation to Neuroengineering and Medicine

Clara F. López-Leon, Akke Mats Houben, Marc Montalà-Flaquer, Mireia Olives-Verger, Anna-Christina Haeb, and Jordi Soriano

Abstract In this chapter, concepts from physics of complex systems are applied to understand the collective behavior of neuronal cultures, in which an ensemble of initially isolated neurons self-organize and connect to one another to exhibit spontaneous activity with non-trivial dynamics. These cultures are an important experimental model since they allow to study, and even act on, the ensemble of neurons in a controlled manner. Specifically, here we investigate how spatial constraints in the arrangement of neurons and connections on the substrate they grow orchestrate the collective behavior of the emerging network, for instance by shaping the richness of spontaneous activity patterns or the velocity of propagating fronts. We also investigate the importance of the balance between excitatory and inhibitory connections in modulating the repertoire of activity patterns. Given the controllability and versatility of neuronal cultures, we also shed light on the use of numerical models to illustrate their potential in reproducing experimental behavior with relatively simple concepts, reveal the key factors influencing network dynamics, and make predictions. Finally, we discuss the medical applications of neuronal cultures together with their limitations and drawbacks.

1 Introduction

Physics of complex systems has emerged in the last decades as a powerful branch of knowledge to model and predict both the natural world and human-made technologies. From social interactions to the brain and internet, complex systems deal with the understanding of how an ensemble of interacting units are able to exhibit a collective behavior that is much richer than the sum of its individual components [58]. With the support of powerful tools grounded in dynamical systems [63], data science [80]

Clara F. López-Leon and Akke Mats Houben: These authors contributed equally.

C. F. López-Leon · A. M. Houben · M. Montalà-Flaquer · M. Olives-Verger · A.-C. Haeb · J. Soriano (✉)
Departament de Física de la Matèria Condensada & Institute of Complex Systems (UBICS), Universitat de Barcelona, Barcelona, Spain
e-mail: jordi.soriano@ub.edu

© The Author(s), under exclusive license to Springer Nature Switzerland AG 2025

M. Stich and J. Carballido-Landeira (eds.), *Nonlinear Dynamics for Biological Systems*, SEMA SIMAI Springer Series 40,
https://doi.org/10.1007/978-3-031-99044-1_3

and complex networks [4], research in complex systems has both revolutionized our view of the world and provided insight into its evolution, playing a fundamental role in understanding problems as diverse as the propagation of epidemics [61], climate change [14], the prognosis of neurological disorders [41, 62], or the vulnerability of the internet [39].

A dream for any scientist working in complex systems is the capacity to *play* with such a system, i.e., to shape and control the interactions among its building blocks as well as to act on the system to dissect the mechanisms that shape its complexity and repertoire of behaviors. This is particularly attractive in biological systems, in which processes at different spatial and temporal scales coexist, from molecular interactions to multicellular communication. One of such *model complex systems* is a neuronal culture, i.e., the growing of neurons in a controlled, in vitro environment.

Surprisingly, neuronal cultures were already developed about 100 years ago [42], first to verify that neurons were the building blocks of a nervous system, then to understand their intrinsic dynamical behavior, and finally to investigate their capacity to connect to one another and shape intricate information processing networks [78]. Along this century of neuronal cultures, the scientific community has acknowledged their importance as a toolbox to understand the brain and its dysfunctions. And it is precisely the fact that one can act on neuronal cultures what has made them so attractive. For instance, various electronic and optical technologies allow to monitor the evolution of a neuronal culture from a community of isolated cells to a networked assembly with rich topological and dynamical characteristics, providing invaluable insight in aspects such as self-organization, spontaneous activity, the emergence of rich non-random network traits, critical behavior, computation, or the repertoire of activity patterns [3, 8, 47]. Additionally, with the advent of microfabrication technologies and bioengineering, neuronal cultures have been used to explore in detail the relationship between structural connectivity and collective dynamics, the quest towards mimicking brain-like circuits *vitro*, or the vulnerability of neuronal networks to perturbations, damage or disease [32, 36, 68, 72]. The latter is related to a wealth of multidisciplinary efforts aimed at understanding and treating neurological disorders, in which neuronal cultures provide an environment to explore both the mechanisms behind network alterations and the goodness of therapeutic interventions [17, 18, 53].

In this context of neuronal cultures as a model system for emerging complex behavior, here we present an overview of experimental, analytical, and numerical explorations that we have conducted in the recent years to understand the behavior of biological neuronal networks. In the sections below, we will provide first examples of different neuronal culture preparations and the tools that we used to analyze their dynamics. We will then introduce numerical simulations to illustrate their capacity to model and predict experimental results. Finally, we will illustrate the medical application of neuronal cultures by comparing their dynamical behavior in physiological and diseased states. We believe that our work can be of inspiration for those scientists working in multidisciplinary research fields.

2 Preparation of Neuronal Cultures and Their Analysis

Here we provide a description of the techniques used to prepare neuronal cultures under different configurations and the data analysis pipeline implemented to extract the data.

2.1 Preparation of Neuronal Cultures and Their Analysis

Standard neuronal cultures are prepared by extracting neural tissue from either rat or mouse embryonic brains. Mechanical dissociation is then performed to obtain single neural cells through gentle pipetting, and finally plating the neural cells on a substrate precoated with adhesive proteins [42, 59]. Adequate culture media in combination with incubation protocols support neural survival and differentiation, facilitating the development of neurites (axons and dendrites) that shape a de novo neuronal network with rich spontaneous activity. It must be noted that the initial neural tissue contains both neurons and neural progenitors. The latter differentiate into glial cells in the following days stimulated by growth factors present in the culture medium, ultimately leading to a network that contains both neurons and glia. These glial cells play an important role in nurturing the neurons and regulating synaptic connectivity and plasticity [1, 56], but their behavior as well as neuron-glia interactions are normally disregarded in those studies in which the central focus of attention is the neuronal network, as in our case.

Neuronal networks derived from embryonic brains are termed *primary neuronal cultures* since they maintain the characteristics of their tissue of origin, i.e., the neuronal specificity and relative population ratios. Specificity means that neuronal types, such as excitatory pyramidal neurons or inhibitory interneurons, as well as the types of synapses (excitatory or inhibitory), are retained when the primary culture forms. For instance, cortical tissue leads to primary cultures with typically 80% excitatory neurons and 20% inhibitory ones, whereas hippocampal tissue provides 70% excitatory and 30% inhibitory neurons [60]. Also, a neuron of a particular type establishes connections of the same type, so that an excitatory neuron will perform the same excitatory action on all neurons it connects to (Dale's principle [35]). The maintenance of cellular specificity is what makes primary cultures a gold standard in many research fields, from physics to medicine, since it ensures reproducibility in the preparation of the cultures and a similar development of the network across experimental realizations.

An alternative approach to obtain neuronal cultures is through *induced pluripotent stem cells* (iPSCs), normally from human origin (hiPSCs), in which somatic cells are reprogrammed to a pluripotent stage via specific transcription factors and, later, differentiated into neurons and other nerve cells [46, 57, 66]. The advantage of hiPSCs-derived neuronal cultures is that they preserve their intrinsic human traits [10,

81] and avoid the use of animals for the harvesting of cells. Additionally, hiPSCs-derived cultures allow to compare healthy neuronal networks with those derived from hiPSCs harboring genetic anomalies that are associated to neurological diseases [38], such as Huntington's [9], Sanfilippo [6] or some forms or Parkinson's [7], thus allowing to study disease progression and the design of therapeutic targets in vitro.

Despite the relevance of hiPSCs technology, with a clear leadership in cell culturing for the years to come, their preparation is long and difficult, and often lead to neuronal cultures with a large variety of cell types. This richness can be viewed as a drawback, particularly in the context of neuroengineering and physics of complex systems, where a stability in cellular types allows for a better modeling and understanding of emerging network phenomena. Thus, in the results that we provide here, we concentrated solely on data obtained from primary neuronal cultures.

Figure 1 provides examples of the typical neuronal cultures prepared in our laboratory together with illustrative brightfield images. Three distinct preparations are considered, namely a uniform distribution of neurons on a surface (*homogeneous culture*), an ensemble of interconnected groups of neurons (*aggregated*), and an engineered neuronal population whose connectivity is guided by means of a patterned substrate (*topographical*). The standard culture design corresponds to neurons plated

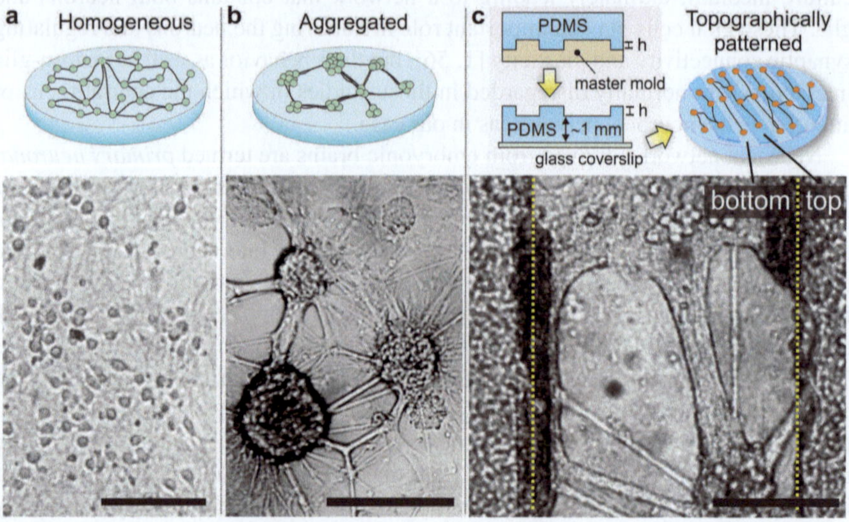

Fig. 1 Representative neuronal cultures with the three distinct organizational features, namely homogeneous, aggregated, and topographically patterned. Top, schematic representation of the different culture designs. Bottom, brightfield images illustrating each kind of preparation. **a** In the homogeneous case, neurons are plated uniformly on a flat surface. Neurons slightly group as they mature, but single cell bodies can be clearly detected. **b** In the aggregated case, neurons cluster in small, sphere-like islands connected to one another in which single neurons cannot be resolved. **c** In the topographical case, neurons are plated on a two-level PDMS substrate that constraints or guides neuronal connectivity. The dashed yellow line indicates the separation between topographies. Black scale bars are 100 μm

homogeneously on a transparent flat surface, e.g., glass, or silicone elastomers such as polydimethylsiloxane (PDMS). In these cultures, individual cell bodies can be well identified or they shape a uniform layer across the culture, although some small aggregation caused by cell motility and axonal tensile forces may occur [54]. We used such a system to investigate the emergence of spontaneous activity [27, 47], its alteration through the modulation of the excitatory-inhibitory balance [74], connectivity [34, 60], and resilience to damage [2]. In contrast to homogeneous cultures, aggregated neuronal cultures are those in which neurons cluster in compact islands about 100 μm across connected to one another by bundles of axons. These cultures are obtained by either plating neurons on small precoated drops or simply on a bare surface, fostering neuronal mobility and their grouping [29, 54]. We used these cultures to study self-organized modular networks [70] and resilience to damage [69, 71]. Finally, topographical cultures allow to combine characteristics of the two above designs. These cultures are prepared by casting a two-level PDMS topographical pattern from an engineered master mold, and neurons are then plated on the surface of the pattern. The topography is shaped as parallel tracks, but arbitrary designs can be introduced. Neurons grow by combining a homogeneous and aggregated organization, but the presence of the pattern breaks the isotropy in connectivity and neurons connect more strongly along the tracks than transverse to them. We used this design, and conceptually similar ideas using chemical confinement instead of topographical, to investigate the relationship between activity and connectivity as well as the richness in spontaneous activity patterns [44, 77].

2.2 *Calcium Imaging and Data Analysis*

Regardless of the configuration of the neuronal cultures, we investigated their behavior through the analysis of spontaneous activity recorded by means of *calcium imaging* [25] (Fig. 2). This technique is based on the indirect detection of neuronal activations (action potentials or *spikes*) associated to the large increase in intracellular calcium concentration upon the opening of voltage-gated calcium channels. By using small organic compounds with fluorescent properties when bound to calcium, the behavior of a large population of neurons can be monitored with a temporal resolution in the range of 10–20 ms. Additionally, depending on the setting of the optical system, activity can be monitored either at a cellular scale (Fig. 2a, b) or across small regions of interest (ROIs) that may contain few neurons (Fig. 2c). In general, the small organic compounds (or calcium indicators) can be either chemical [50] or genetically encoded [37]. The former are directly loaded into the cells by increasing their permeability through solvents, while the latter are incorporated into the cells through viral transfection. In the experiments that we report here we used the genetically encoded calcium indicator (GECI) GCaMP6s in combination with adenso-associated viruses. A general advantage of GECIs is that the same network can be studied along several days, e.g., to quantify the evolution of a neuronal network [10, 44], or before and after a perturbation, e.g., to investigate the impact of a perturbation [2].

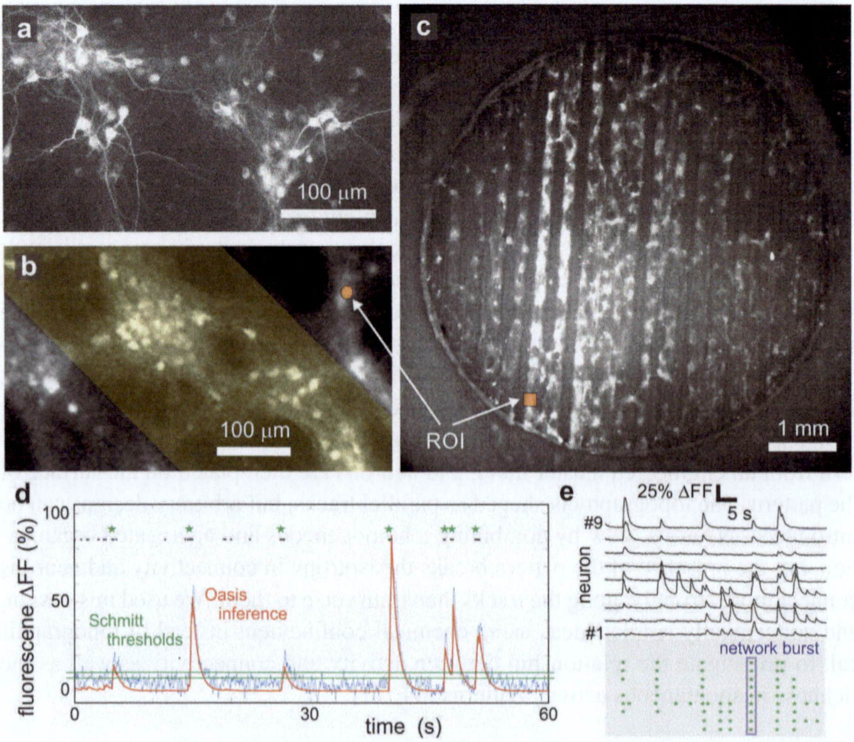

Fig. 2 Data acquisition and analysis. **a** Representative fluorescence images of a homogeneous neuronal culture with slight aggregation. Bright quasi-circular objects are active neurons. **b** Detail of a fluorescence image of a culture grown on a PDMS substrate. The yellow area indicates the bottom of the PDMS topography. **c** Corresponding culture imaged entirely, i.e., in a field of view that covers the whole 6 mm diameter culture. The orange spots mark regions of interest (ROIs) that encompass either a single neuron or a small region of the large 6 mm cultures. **d** Data analysis consists in extracting the average fluorescence within each ROI of a culture as a function of time, and then inferring the activation times of the neurons (black dots or green stars). **e** The fluorescence data of all ROIs (top) is transformed into a raster plot that portrays the detected neuronal activations as a function of time (bottom). A 'network burst' is defined as the coherent activation of a large fraction of the network in a short time window (blue outline)

A typical calcium imaging experiment in our laboratory consists of the recording of spontaneous activity for about 10–30 min at typically 50 images/s (20 ms acquisition interval), followed by the analysis of the obtained fluorescence data. For that, the images were first analyzed to identify the ROIs (either single cells or small areas) and then extract the average fluorescence signal within each ROI (Fig. 2d, blue trace). This signal was then normalized as $\Delta F F_i(t) \equiv (f_i(t) - f_{i,0})/f_{i,0}$, where $f_i(t)$ is the raw fluorescence signal of neuron i and $f_{i,0}$ its background level, i.e., in the absence of activity. The $\Delta F F(t)$ data was then examined to extract neuronal activations or spikes. In the data present here we used the Schmitt trigger method [24], which accepts a spike whenever the fluorescence signal exceeds preset thresholds (green lines and stars in Fig. 2d). Other methods such as Oasis [21] (red calcium signal and black dots) use biophysical information such as the indicator kinetics to predict the spike trains. Although the Oasis method provides higher details on neurons' dynamics, we found that the Schmitt method was more adequate since it just reveals the moment of a neuronal activation, regardless of concatenated repetitions, which we found particularly suited to study traits such as the richness in spatiotemporal activity patterns or their velocity of propagation. Finally, for a given set of ROIs and fluorescence traces, the final data takes the form of a raster plot (Fig. 2e), from which further analysis can be carried out. We notice that, in this raster plot representation, we can easily visualize collective neuronal events or *network bursts* as the quasi-synchronous activation of a large number of neurons in a small time window (blue dashed outline).

3 Experimental Results

Taking advantage of different neuroengineering resources to constrain the position and connectivity of neurons, here we compare the behavior of homogeneous, aggregated, and PDMS topographical patterns, and show that collective dynamics, as well as the velocity of propagating fronts, are greatly influenced by two main factors, namely the underpinned network layout imprinted by the spatial constraints and the balance between excitatory and inhibitory connections.

3.1 Spatial Constraints Shape Network Dynamics

Figure 3a compares the collective dynamics of three configurations. All of them correspond to cultures whose spontaneous activity was recorded at day in vitro (DIV) 12 for 10 min. The first configuration (left), corresponds to a neuronal culture with about 700 neurons that were homogeneously distributed and firmly attached to a flat surface by means of adhesive proteins, shaping a neuronal network as in Fig. 1a in which potentially all neurons connect to others in their neighborhood in an isotropic manner. We explored such a configuration in detail in previous studies [44, 47, 73, 74] and observed that such a homogeneous distribution favors a dynamical behavior characterized by highly coherent, quasi-synchronous collective activations in which all neurons activate in a short time window (*network bursts*) and remain silent in between bursts. Such a strong coherence is well captured by the 'global network activity' (GNA), which portrays the fraction of the neurons in the network that coactivate in a short time window of typically 200 ms [44]. Thus, for the homogeneous configuration, the network is locked in a repetitive all-or-none behavior that confers it a very poor repertoire of dynamical states.

When spatial patterning of neurons was introduced, however, collective dynamics became much richer. Such a spatial patterning was achieved by preparing aggregated cultures as in Fig. 1b, in which neurons clustered in small islands connected to one another. This configuration was realized by either growing neurons on bare glass (as the case shown in the figure), fostering their aggregation and shaping *self-organized modular networks* [70, 71], or by imprinting neurons in predefined locations in an engineered modular way [55, 77]. In either case, the neuronal aggregates effectually shaped highly intra-connected units that were typically weakly inter-connected, or with varying degrees of connectivity that could be tuned upon engineering [77]. In our case we prepared a self-organized network containing about 170 aggregates and monitored their spontaneous activity, and observed a collective dynamics that was clearly much more varied (Fig. 3a, center). The aggregates indeed coactivated in groups of different sizes, from about 15–100% of the network, leading to a GNA plot where peaks of different amplitudes coexisted. We ascribe the variability in collective activity to the randomness in the strength and number of connections between aggregate pairs. This variability is supported by previous studies in which the repertoire of coactivation patterns across the network changed upon alteration of the connectivity strength between aggregates through chemical action [11].

An alternative approach to imprint spatial patterning in a neuronal culture consists in using PDMS topographical obstacles that restrict the position of neurons and constrain their connectivity. An enlightening approach is the use of PDMS topographical tracks (Figs. 1a and 2c), in which the neurons grow either at the top or at the bottom of the tracks and develop connections that are much more abundant along tracks than across them [44]. In such a configuration, we observed that the collective dynamics of the culture was also substantially enriched (Fig. 3a, right), with an abundant variety of coactivation patterns that comprised from a single track to the entire network. We ascribe this richness to the fact that the tracks themselves act as modules, with neurons strongly connecting within a track but weakly connecting to neurons in other tracks. Thus, we conclude that a rich repertoire in activity is related to, or clearly favored by, a structural connectivity with modular characteristics, a feature that the homogeneous network lacks.

To further explore the differences between the three culture configurations, we studied the spatiotemporal propagation of spontaneous activity (Fig. 3b). Such a study was carried out by first identifying, in the GNA profile, each peak i with amplitude $\Gamma_{t,i}$ and retrieving the activation times and spatial coordinates of the neurons that participated in it. We discarded peaks' amplitudes $\Gamma_t < 0.1$ since they could often be confused with sporadic network activations. Then, the neurons that initiated activity were identified and the structure of the spatiotemporal front was plotted as a heat map. As shown in Fig. 3b, the homogeneous cultures exhibited a propagation consistent with a circular front, and with velocities in the range 40–50 mm/s. These values were determined as a linear fit of the distance of each ROI to the origin of activity as a function of its activation time (Fig. 3c). The quasi-circular front propagation suggests an isotropic connectivity combined with a characteristic length scale for neurons to connect to one another. For aggregates, propagation was characterized by chained activations that often led to serpent-like excursions across the culture,

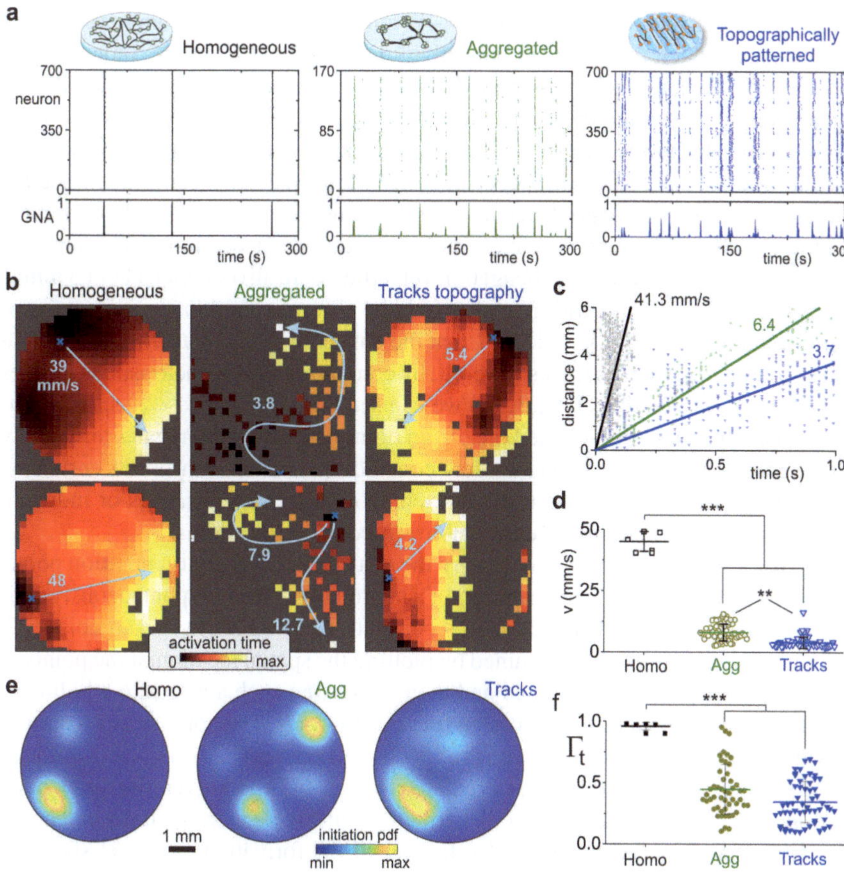

Fig. 3 Dynamics in different neuronal cultures at day in vitro 12. **a** Raster plots and global network activity (GNA) of spontaneous activity in 6 mm diameter neuronal cultures grown on: a flat surface with a homogeneous distribution of neurons (black, left), a flat surface with neuronal aggregates (green, center), and a PDMS topographical surface shaped as parallel tracks that guide neuronal connections along them (blue, right). For clarity, only the first 5 min of recording (of the total 10 min) are shown in the raster plots. **b** Illustrative spatiotemporal fronts of activity propagation for the three networks. Velocities are much lower for the non-homogeneous configurations. Black-yellow maps show the relative activation time of each ROI. Blue crosses mark the origin of activity. Scale bar is 1 mm. **c** Extraction of the characteristic velocities of propagating fronts as a linear fit (color lines) of the distance of ROI to the origin of activity as a function of ROIs' activation time. **d** Distribution of measured characteristic velocities for the three configurations. The distributions are significantly shifted toward lower values as more spatial constraints are imposed on the connectivity. **e** Spatial distribution of activity initiation, which is more spread for stronger physical confinement. Black dots are the initiation points of measured network bursts. **f** Distribution of collective events sizes Γ_t (peaks of the GNA plots), which are significantly more spread for the aggregated and topographical cases, indicating a rich variability in collective activity. ***: $p < 0.001$; **: $p < 0.01$, Student's t-test

and with typical velocities in the range 4–15 mm/s, much smaller than in the homogeneous case which reveals that connectivity is strongly short range. Indeed, if we take a typical integration time of 5 ms for the incoming activity in an aggregate and a characteristic inter-aggregate distance of 50 μm, we obtain a typical velocity of 10 mm/s, similar to the measured values. For tracks, activity propagation was not circular-like but more varied in space, and often the propagating front was embedded in regions that were strongly faceted along the direction of the tracks (bottom-right front in Fig. 3b), indicating that inter-track connectivity shaped in great manner the structure of the front. Typical propagation velocities were also smaller, about 5 mm/s. Interestingly, the velocity within tracks could be also measured in some fronts, providing about 20 mm/s [44], indicating that the connectivity along tracks was about 8 times stronger than across tracks. Overall, by comparing the characteristic velocities in the three configurations (Fig. 3c, d) we can see that they decrease as spatial constraints become more pronounced, i.e., from homogeneous to aggregates and finally to topographical tracks. We can therefore conclude that spatial constraints enrich the connectivity layout among neurons, shortening connectivity distances or inducing strong anisotropies that translate into more varied activity patterns.

We also found it interesting to observe the distribution of activity initiation points (Fig. 3e) which is related to the capacity of a neuronal network to spontaneously activate due to the interplay between spatial correlations in connectivity and noise [27, 47]. These distributions were obtained by plotting the spatial location of the points in which each detected front initiated and then converting such a map into a bidimensional initiation probability distribution function [44, 47]. The homogeneous culture provided a highly focalized initiation (yellow spot), i.e., most network bursts initiated in the same spot, a result that was predicted in Ref. [47]. However, the aggregated and topographical cultures exhibited distributions that were more spatially extended, indicating that spatial constraints also facilitated the formation of 'sweet spots' of stronger local connectivity that facilitated activity initiation.

We note that the distribution of GNA peaks Γ_t (Fig. 3f) serves as a basis to quantify the richness in collective activity, i.e., the abundance of coactivation events of different sizes, allowing to provide a single descriptor, Θ_{GNA}, to describe such richness. This descriptor is computed as follows. Following Refs. [77, 79], we first pooled together all GNA peaks' amplitudes Γ_t, which vary between 0 and 1, and then computed the corresponding distribution $p(\Gamma_t)$. Finally, Θ_{GNA} was obtained as

$$\Theta_{\text{GNA}} = 1 - \frac{m}{2(m-1)} \sum_{\mu=1}^{m} \left| p_\mu(\Gamma_t) - \frac{1}{m} \right|, \tag{1}$$

where $|\cdot|$ denotes the absolute value and $m = 20$ is the number of bins used for estimating the distributions. Θ_{GNA} takes values between 0 and 1. The larger Θ_{GNA} is, the more varied are the Γ_t values. From the data of Fig. 3f, we obtained Θ_{GNA} values of 0.05, 0.55 and 0.51 for homogeneous, aggregated and topographical configurations, respectively.

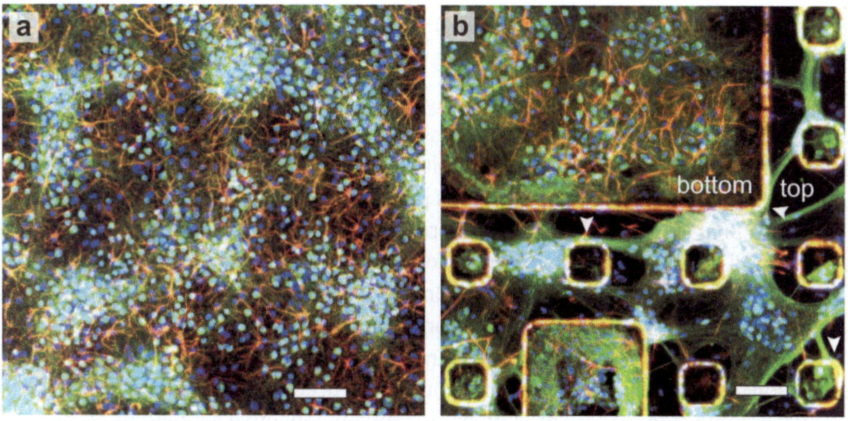

Fig. 4 Immunostaining of neuronal cultures. **a** Homogeneous cultures, in which neurons cover uniformly the substrate with some aggregation. Connections spread out in all directions. **b** PDMS topographically-patterned culture, in which neurons often appear confined at the bottom of PDMS molds and connections follow the edges of the obstacles (white arrowheads). In both images: magenta corresponds to DAPI and indicates cell nuclei; green corresponds to MapII and identifies neurites; red corresponds to GFAP and identifies glial cells. Scale bar is 100 μm

Finally, we would like to emphasize that the observed richness in activity for topographical cultures can be ascribed to connectivity features at a microscopic level by using immunostaining or other connectivity-labeling techniques. Different studies have used this strategy to measure average axonal distributions and lengths [15, 16] or the interaction of axons with obstacles [43, 51]. In the context of our study, Fig. 4 provides an example of immunostained cultures for a homogeneous configuration and a PDMS-topographical one. The latter corresponds to a PDMS layer (labelled 'top') patterned with squared valleys of different dimensions ('bottom'). The homogeneous configuration (Fig. 4a) shapes a quasi-uniform carpet of neurons (magenta) and glia (red) in which neuronal processes (green) spread in all directions. For the topographical case (Fig. 4b), the distribution of neurons is more varied, combining homogeneous distributions (e.g., left valley) within aggregated ones, while neuronal processes clearly tend to group in thick bundles that often follow the contours of the topography. Indeed, as highlighted by the arrowheads, neurites tend to bend at the edges of the topography or follow it. This behavior illustrates the capacity of PDMS obstacles to guide connections and promote anisotropies. A similar immunostaining analysis for the tracks' topography is shown in Ref. [44] and provides evidence of the strong anisotropy induced by this configuration, with most processes following the orientation of the tracks and few of them crossing the tracks.

3.2 Importance of the Excitatory-Inhibitory Balance

To complete the experimental investigations, we explored the impact of the excitatory-inhibitory balance in collective dynamics. Our cortical cultures contain 80% excitatory synapses and 20% inhibitory ones. Since inhibition plays an important role in regulating network dynamics [52, 65], we inquired whether an alteration of this balance substantially altered the richness of spontaneous activity in neurons grown on PDMS topographical tracks. For that, we recorded activity for 10 min with both excitatory and inhibitory connections active, then inhibition was blocked with 40 μM bicuculline [60, 74] and we recorded activity again for another 10 min.

Figure 5a compares the raster plots and global network activity before and after inhibition blockade. Strikingly, the rich variability in collective events was practically lost when inhibition was silent, with practically all collective activations comprising 100% of the network. We note that full network activation upon bursting was not achieved, i.e., the culture did not behave as a homogeneous one, indicating the PDMS topography had still the capacity to enrich collective dynamics. However, as shown in Fig. 5b, the distribution of GNA amplitudes Γ_t switched from substantially broad to highly peaked toward 1 when inhibition was silent, with a corresponding an overall richness that abruptly changed from $\Theta_{\text{GNA}}^{\text{E+I}} \simeq 0.51$ to $\Theta_{\text{GNA}}^{\text{E-only}} \simeq 0.12$. This result strongly supports the idea that inhibitory connections play a key role in shaping the dynamic behavior of neuronal circuits.

Finally, we also computed the characteristic propagation velocities of network bursts' fronts (Fig. 5c). A blockade of inhibition effectively increases overall excitation [60] and thus enhances activity propagation [34]. Therefore, as expected, we observed a significant increase, by a remarkable factor 4, of propagation velocity, although the measured values are still below that of the homogeneous cultures, which indicates again that the connectivity traits shaped by the PDMS topography cannot be completely erased.

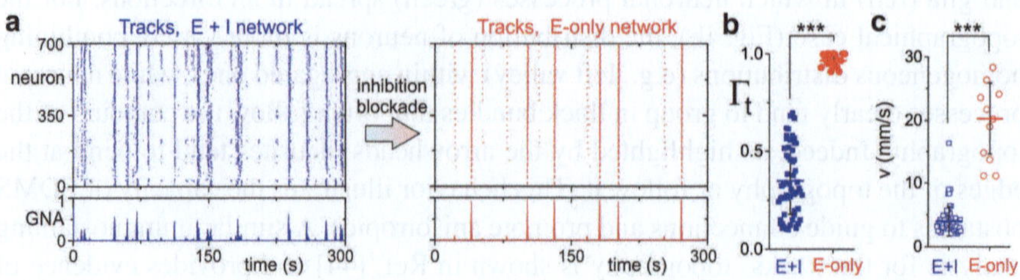

Fig. 5 Impact of the blockage of inhibition on neuronal culture dynamics. **a** Raster plots and global network activity (GNA) of spontaneous activity on a neuronal culture grown on topographical tracks, comparing the condition in which both excitation and inhibition are active (left) with the one in which inhibitory synapses have been fully blocked with bicuculline (right). **b** The size of collective events Γ_t significantly increases upon inhibition blockage. **c** The velocity of propagating fronts significantly increases, by 4 times, upon inhibition blockage. ***: $p < 0.001$, Student's t-test

4 In Silico Modeling

The above results illustrate that neuronal cultures are unique laboratories to investigate the relationship between network connectivity and emergent dynamics. Given the relative simplicity of the cultures, provided by their small size and uniformity in cell types, experimental studies can be combined with numerical models to fully comprehend this relationship. Numerical models take indeed advantage of this simplicity to consider scenarios in which collective activity is the main observable, and where the coupling of $\sim$1000 neurons through a complex network suffice to capture key experimental observations without the need of incorporating a wealth of microscopic details. The scalability of numerical models, in which new features can be gradually incorporated when needed, also gives room for a wealth of explorations that can be then used to plan new experiments, uncover hidden mechanisms, or make predictions. For instance, we used in silico modeling to understand in detail the mechanisms that shape spontaneous activity and their relation to the intrinsic nonlinear properties of neurons [12], noise [27, 47], spatial embedding [28, 73], connectivity guidance [40], and damage [13, 27].

To illustrate the capacity of in silico modeling to help understanding experimental behavior, here we investigate the combined influence of connection distance and connection strength on the dynamics of neuronal cultures. We used numerical simulations of $N \approx 3000$ spiking neurons with different average axon length $\langle \ell \rangle$ and excitatory connection strengths g_{AMPA}.

4.1 Neuronal Model

Each neuron was modeled as a regular spiking Izhikevich model neuron [30] for which the membrane potential v_i and adaptation current u_i of a neuron i evolved according to

$$\begin{aligned} \frac{dv_i}{dt} &= 0.04v_i^2 + 5v_i + 140 - u_i + I_i^{\text{syn}} + \sigma\eta_i(t), \\ \frac{du_i}{dt} &= a(bv_i - u_i), \end{aligned} \tag{2}$$

where I_i^{syn} is the synaptic input received from other neurons, $\eta_i(t)$ a normally distributed white noise process, $\langle \eta_i(t)\eta_i(t')\rangle = \delta(t-t')$, a and b model parameters, and σ the noise magnitude. Dynamically, whenever the membrane potential of neuron i crosses a preset threshold v_{th} the neuron is said to spike and the membrane potential v_i and adaptation current u_i are reset as

$$v_i > v_{th} \implies \begin{cases} v_i \leftarrow v_{\text{reset}} \\ u_i \leftarrow u_i + d, \end{cases} \tag{3}$$

Table 1 Parameters used for the numerical simulations shown in this article

Parameter	Value	Parameter	Value
a	0.02	g_{AMPA}	4 mV
b	0.2	g_{GABA}	8 mV
d	6.5	τ_P	10 ms
v_{reset}	−65 mV	δ_P	1
v_{th}	30 mV	τ_R	8×10^3 ms
σ	2.25 mV^2s	δ_R	0.8

with v_{reset} and d constants. We note that the parameters a, b, v_{reset}, and d can be adjusted to reflect different neuronal behaviors [30, 31]. Indeed, a describes the recovery time scale of the adaptation current u, b its sensitivity to subthreshold fluctuations of the membrane potential v, v_{reset} the after-spike reset value of v, and d the after-spike reset of u. The larger d, the slower the recovery of the neuron after a spiking event.

The synaptic inputs

$$I_i^{\text{syn}} = \sum_j g_j W_{ij} P_j(t) \tag{4}$$

that a neuron receives depend on the connectivity matrix $\mathbf{W}$, the construction of which is explained below. The synaptic strength g_i depends on the pre-synaptic neuron type, with $g_i = g_{\text{AMPA}}$ for excitatory and $g_i = g_{\text{GABA}}$ for inhibitory neurons.

The synaptic potentials $P_i(t)$ follow exponential decay with time-constant τ_P. Upon firing of the pre-synaptic neuron i the synaptic potential P_i is increased with a factor depending on a variable $R_i(t)$ that captures the number of available neurotransmitter vesicles for release. This variable follows

$$\tau_R \frac{dR_i}{dt} = 1 - R_i, \tag{5}$$

and P_i, R_i are updated following a spike of neuron i as

$$v_i > v_{th} \implies \begin{cases} P_i \leftarrow P_i + \delta_P R_i(t) P_i(t) \\ R_i \leftarrow R_i - \delta_R R_i(t), \end{cases} \tag{6}$$

The parameters used for the simulations are reported in Table 1. The Izhikevich model neuron was chosen since it provides a good balance between biophysical detail and computational efficiency [30, 31]. We carried out some exploratory simulations (not shown) in which the dynamics of the membrane potential v was substituted by other model neurons, such as simpler linear integrate-and-fire neurons or more complex FitzHugh-Nagumo [20, 45] models. We observed that an alteration of the membrane potential dynamics while maintaining the rest of parameters fixed—especially the synaptic dynamics—did not result in qualitatively different results. Indeed, the main necessary ingredients for a neuronal network to exhibit complex behavior are excitable elements coupled with synapses that are depressed on an intermediate time-scale and recover on a long time-scale (τ_R), with considerable freedom to choose the specifics of the internal neuronal dynamics.

4.2 Network Construction

The connectivity matrix **W** was constructed by modeling axon growth following [47]. In the basic construction, $N = 3000$ neurons were uniformly placed on a circular surface of diameter $D = 3$ mm. Each neuron had a size of $r_{\text{soma}} = 7.5$ µm, and neurons were placed in a non-overlapping manner. Next, from each neuron i an axon of length ℓ_i was 'grown' by concatenating line segments of length $\Delta\ell = 10$ µm with each segment placed with an angle $\varphi_n \sim \mathcal{N}(0, 0.1)$ radians relative to the angle of the previous segment φ_{n-1}. Once a line segment of the axon of neuron i was placed within a distance of $r_j^{\text{denrite}} \sim \mathcal{N}(150, 20)$ µm of another neuron j, a connection from i to j was established with a probability $\alpha = 0.5$.

We have seen that neuronal cultures grown on a flat surface tend to display highly coherent activity, and that by using PDMS topographical patterns this coherence was broken and richer dynamics emerged. To model numerically the presence of PDMS tracks, vertical bands 200 µm wide and spaced 300 µm apart were placed on the surface, as sketched in Fig. 6a. Once an axon segment was placed across the border of one of these obstacles, it remained there with a certain probability; otherwise, it was replaced by a line segment parallel to the obstacle wall. The crossing probability depended on the direction of growth. Axons growing from the bottom to the top of the PDMS obstacle had a probability of P_{up} to remain, and a $1 - P_{\text{up}}$ probability to be replaced. Growing from the PDMS downwards was successful with a probability $P_{\text{down}} > P_{\text{up}}$.

The total axon lengths ℓ_i were drawn from a Rayleigh distribution with average length $\langle\ell\rangle$, a parameter that, as described next, was varied to explore different conditions of the simulations. The Rayleigh distribution was derived from experimental observations [16] and reads

$$p(\ell) = \frac{\ell}{\sigma_\ell^2} \exp\left(\frac{-\ell^2}{2\sigma_\ell^2}\right), \tag{7}$$

where $\sigma_\ell^2 = 900$ µm is the variance of the distribution and its value is chosen so that the average axonal length is $\langle\ell\rangle \simeq 1.1$ mm for the experimental data.

An example of a constructed network is shown in Fig. 6b. Once the network was set up, 80% of synapses were set as excitatory and the rest as inhibitory.

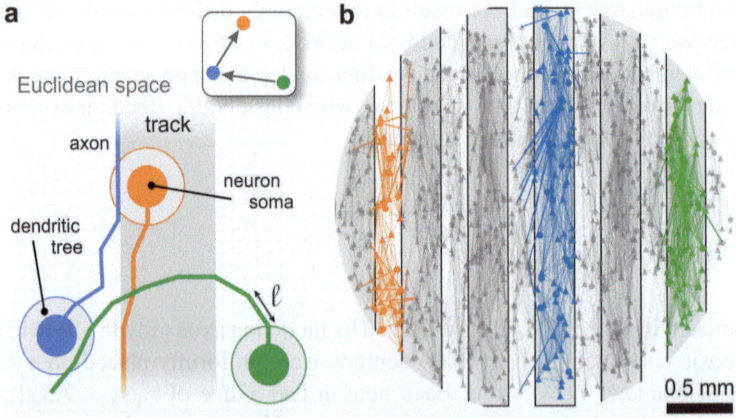

Fig. 6 Construction of in silico networks. **a** Neurons (solid discs) are placed on a bi-dimensional surface and neurons grown by concatenating segments of length $\Delta\ell$. Axons have a given probability to either follow the tracks' obstacles (blue and orange) or cross the tracks towards neighboring areas (green). Whenever an axon of neuron i intersects the dendritic tree (color circles) of neuron j, a connection $i \rightarrow j$ is established with probability 0.5 (top-right sketch). **b** Final network grown on a 3 mm diameter circular area. Up-triangles and circles indicate excitatory and inhibitory neurons, respectively, and lines connections. Colors indicate groups of neurons strongly connected within their physical area, and their connections towards neighboring tracks

4.3 *Results: Connectivity and Spatial Constraints Shape Activity*

We explored the behavior of network collective activity for different values of excitatory connectivity strengths and average axonal lengths. For clarity of discussion, excitatory strength was indicated as $g^*_{\mathrm{AMPA}} = g_{\mathrm{AMPA}}/g_0$, i.e., relative to standard values used in previous simulations with $g_0 = 4$ mV [27, 47], while average axonal lengths $\langle\ell\rangle$ were expressed as relative to network's spatial diameter, $\delta = \langle\ell\rangle/D$, to better conceptualize the concept of short- and long-range connections [28]. Indeed, for $\delta \ll 1$ neurons connected to their neighbors at the scale of the imprinted topographical tracks, potentially shaping highly spatially localized dynamics, while for $\delta \simeq 1$ neurons could connect to one another across the whole network and the impact of the tracks' patterning was effectively erased.

Figure 7a provides representative raster plots of the most interesting scenarios that can be extracted from the simulations. In scenario (1), connections were weak, and neurons connected within the range of a couple of tracks, leading to a very poor activity in which collective activity was rare and predominantly limited to single tracks. This collective activity can be visualized as either bands of coherently activating neurons in the raster plot or as sharp peaks in the global network activity. In scenario (2), an increase by three-fold in connectivity strength permitted the same locally-coupled neurons to effectively communicate more abundantly with the

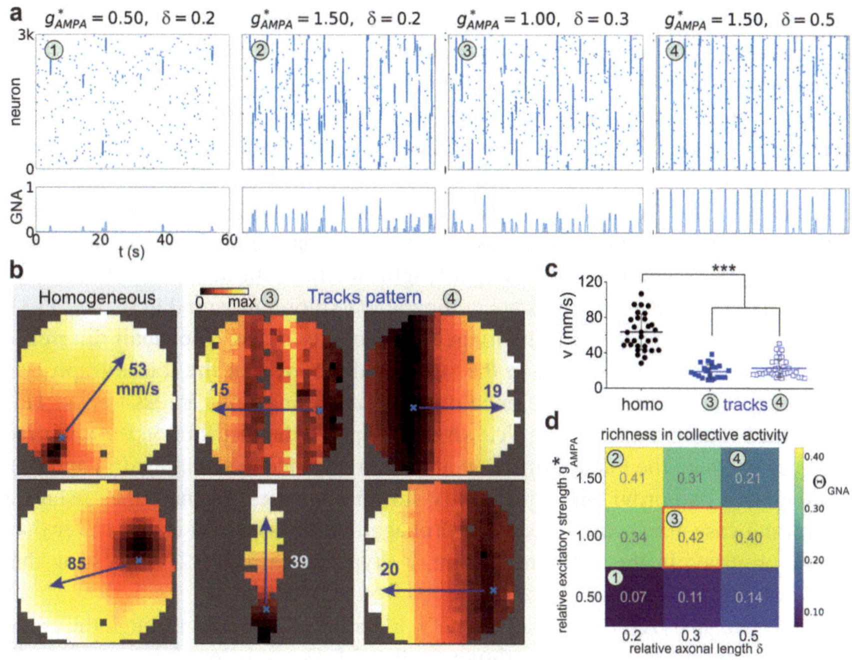

Fig. 7 Neuronal networks in silico. **a** Representative raster plots of spontaneous activity (top) in simulated networks $D = 3$ mm in diameter and a patterned substrate as in Fig. 6b. Global network activity (GNA, bottom) transients from rare to abundant and highly coherent as excitatory connectivity g^*_{AMPA} and relative axonal lengths $\delta = \langle \ell \rangle / D$ grow. **b** Example of spatiotemporal fronts and determined velocity of propagation. The first column shows the reference condition of a flat, unpatterned substrate. The other columns show examples for two of the conditions shown above. **c** Distribution of propagation velocities, comparing the flat condition and the patterned one. ***: $p < 0.001$, Student's t-test. **d** Richness in collective activity, which is maximal for moderate values of connectivity strengths and relative axonal lengths (red square)

neighbors, shaping a stronger and richer collective dynamics in which coactivations of different sizes occurred, from single tracks to the whole network, as evinced by the variability in peaks' amplitudes in the global network activity. Scenario (3) led to a similar behavior, but here a weaker connectivity strength was compensated by slightly larger average axonal lengths. Finally, in scenario (4), both connectivity strengths and axonal lengths were large and the network repeatedly activated in a quasi-synchronous manner.

Representative spatiotemporal fronts for scenarios (3) and (4) are provided in Fig. 7b. They are compared with the behavior of a network in which neurons are placed on a flat surface without patterning, and with parameters identical to scenario (4) ($g^*_{\text{AMPA}} = 1.5$, $\langle \ell \rangle = 1.5$ mm and $\delta = 0.5$). The flat construction leads to quasi-circular fronts that propagate fast at about 50–100 mm/s. For patterning, scenario (3) shapes fronts of varying spatiotemporal structure, combining slowly propagating waves at about 10–15 mm/s across tracks with waves encompassing only a single track at faster velocities of 35–50 mm/s. This contrast between velocities across tracks or transverse to them was also observed in experiments and reflects the tendency of the tracks' pattern to funnel connections along it and substantially increase average connectivity [44]. Indeed, supported by different theoretical and experimental studies [16, 22, 33], the velocity of a propagating front highly depends on connectivity strength and average axonal length, and therefore the analysis of velocities provides an interesting resource for theoreticians and experimentalists to derive structural properties of network connectivity from a global observable such as the velocity. For scenario (4), the longer lengths facilitate the activation of the entire network. However, the velocities are comparable to those measured in scenario (3), although slightly higher given the longer connectivity range. The small difference in velocities is possibly related to the strong anisotropy that the tracks impose in the system, which limits the presence of abundant connections across tracks. This difference is summarized in Fig. 7c, which compares the distribution of measured velocities between the flat construction and the tracks one in scenarios (3) and (4), with the latter two providing significantly slower velocities. We note that, even for the limiting case $\delta \simeq 1$, the velocity of propagation across tracks is still lower than the one observed in the flat construction, by typically a factor 2, suggesting that spatial constraints are difficult to erase.

Finally, we analyzed the richness in network collective activity through the quantity Θ_{GNA} introduced before. As shown in Fig. 7d, Θ_{GNA} is low for scenarios (1) and (4) since collective activity is either highly random ($p(\Gamma_t)$ peaked towards 0) or strongly locked in a fully coherent state ($p(\Gamma_t)$ peaked towards 1), while scenarios (2) and (3) show the maximum $\Theta_{\mathrm{GNA}} \simeq 0.42$ that evinces a high variability in coactivation patterns, a value that is comparable with the one measured experimentally of 0.51. $\Theta_{\mathrm{GNA}} \simeq 1$ is very difficult to reach, and conceptually would correspond to an equiprobable presence of all coactivation sizes Γ_t. The network construction in which neurons grow on a flat surface leads to $\Theta_{\mathrm{GNA}} \simeq 0$ since the activity is locked in a highly coherent state.

In conclusion, this numerical study illustrated that relatively easy in silico constructions can be developed that capture the fundamental results of experimental data. Simulations can then be used to investigate the role of different variables, such as noise or excitation-inhibition balance, in global observables such as collective propagation velocities or Θ_{GNA}, allowing to make further predictions on experimental behavior. Also, simulations allow to explore the emergence of fully coherent activity as $\langle \ell \rangle$ grows, mimicking the biological growth of axons during development, and potentially allow to investigate the robustness of the observables as neurons or connections are lost due to damage.

5 Network Dynamics Alterations in the Context of Medicine

In the previous sections, we have shown that the connectivity of a neuronal network as well as the excitation-inhibition balance shape in great manner network's collective dynamic behavior. This observation has motivated the medical community in neuroscience to raise the following question: if neurons and connections experience degradation due to disease, can this degradation be related to changes in collective dynamics?

This question is important in order to design good in vitro models for medicine, particularly to understand how genetic or molecular deficits affect network functionality. However, the use of primary neuronal cultures to answer this question is not as simple as it seems. The first reason is that, as we reported in previous experimental and numerical studies, neuronal cultures exhibit a strong resilience to damage [2, 27, 69] combined with overt homeostatic plasticity [11, 69]. The latter is associated to the capacity of the neuronal cultures to strengthen or weaken synaptic weights to compensate for perturbations, effectually maintaining the network within stable levels of collective activity [75]. For instance, it has been observed that a reduction in the efficiency of excitatory synapses through AMPA-glutamate receptor antagonists such as CNQX, leads to a gradual strengthening of excitatory synapses to compensate for the loss of excitatory drive [48, 75]. Such homeostatic effects clearly reduce the capacity to detect a network anomaly just from dynamics.

To illustrate the difficulty in linking neuronal affectations due to disease with network-wide alterations, here we provide an example of a neuronal culture that was treated chronically to simulate the action of a pathological agent, in this case altered forms of the protein tau that is associated to a group of diseases known as tauopathies, with Alzheimer's disease being the most prominent one [23]. These diseases are characterized by the abnormal accumulation of tau, a protein that in healthy physiological conditions plays a fundamental role in modulating neuronal microtubule dynamics in the axons of neurons. Upon pathology, however, tau becomes hyperphosphorylated, dissociates from the microtubules, and aggregates into highly insoluble phosphorylated forms [26]. Although tau aggregates are the predominant feature common to all tauopathies, the specific cell types primarily affected, the distribution of these aggregates in different brain regions, as well as the characteristics of the aggregates themselves vary according to the particular disease.

The hypothesis of our in vitro investigation is that the presence of a pathological tau triggers a series of damaging actions that recapitulate those observed in vivo [76], which may include neuronal and axonal loss, disruption of network connectivity, alteration of synaptic plasticity, and a disruption of the excitatory-inhibitory balance. To test our hypothesis, we prepared standard mouse primary cortical cultures shaped as 4 mm diameter homogeneous networks (Fig. 8a) that contained about 700 neurons, chronically treated them with pathological tau (pTau) at day DIV 5 and recorded their spontaneous activity every day up to DIV 16. Identically prepared cultures that served

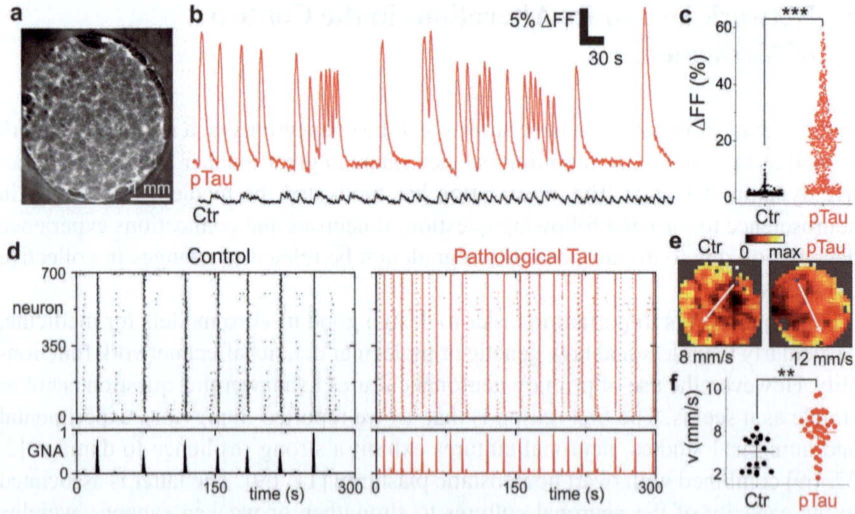

Fig. 8 Disease model in vitro. **a** Fluorescence image of a homogeneous neuronal culture 4 mm in diameter containing about 700 neurons. Two identically prepared cultures were used for the experiments, one cultured in standard conditions (control) and the other one cultured together with pathological Tau (pTau). **b** Network-wide average fluorescence traces for the two conditions. **c** Distribution of neuronal fluorescence amplitudes across the network, with the pTau culture exhibiting significantly larger values. **d** Raster plots of activity (top) and global network activity (bottom). The pTau culture displays a larger number of network bursting events. **e** Example of spatiotemporal fronts, similar in structure but showing a slightly faster velocity for the pTau culture. **f** Distribution of velocities of propagating network bursts, with pTau showing significantly larger velocities. **: $p < 0.01$; ***: $p < 0.001$, Student's t-test

as controls were maintained in standard culture medium and recorded in parallel as the pTau-treated ones.

A first striking difference between the control and pTau cultures is that the latter exhibited a substantially higher amplitude of the fluorescence signal (Fig. 8b), by about 10 fold. This elevated fluorescence suggests that pTau-affected neurons considerably increased their excitability, eliciting much more concatenated action potentials than normal. This hyperexcitability was present across the culture, as shown in Fig. 8c which portrays the distribution of maximum fluorescence amplitudes for all neurons monitored in each condition, and with a statistically significant difference. To deepen the analysis, we extracted the corresponding raster plots using the Schmitt-trigger method (Fig. 8d) and observed that the pTau-treated culture tended to activate collectively in the form of network bursts more frequently, with about 3.8 bursts/min, much higher than the control case with about 1.4 bursts/min. Finally, an analysis of the spatiotemporal propagation of the bursts' fronts (Fig. 8e) showed that they propagated similarly, but with a faster velocity for the pTau-treated culture. Indeed, as shown in Fig. 8f, the propagation velocity was about $v_{\text{tau}} \simeq 8$ mm/s on average for the pTau-treated culture, significantly higher than the control culture with $v_{\text{ctr}} \simeq 5$ mm/s.

The preliminary conclusion from these measurements at DIV 12 is that pTau-treated cultures were driven toward an over-excited state that increased their spontaneous activity, both at the level of spike emissions and collective activity, as well as the velocity of activity propagation. This idea was supported by Alzheimer-related studies [49, 67] that also reported the emergence of neuronal hyperexcitability. However, although the results seem convincing for these particular measurements at DIV 12, the entire story is not. Firstly, we could not find molecular or synapse-level evidences to accredit the hyperexcitability. And, secondly, we suspect that the hyperexcitability was transient, possibly associated to a temporal strengthening of excitatory connections as a response mechanism to an overall connectivity degradation, as observed in Alzheimer's [64]. In this regard, we observed that the dynamic behavior of pTau-treated cultures was not significantly different from controls at DIV 14 and 16, indicating that possible homeostatic actions were already at play that effectually masked the induced network damage.

To conclude, this example shows how difficult it can be to detect alterations due to disease. They may be present, but the complex interplay between neuronal dynamics, connectivity, and plasticity mechanisms hinders their detection and monitoring. Also, alterations may be clear at a particular timepoint and not others, as in our example, with the timepoint itself possibly dependent on neuronal density, cell types, or the nature and severity of the damaging action. Moreover, we must note that the capacity to detect dynamic alterations depends on the disease model, its mechanisms, and even the design of the experiments. For instance, we were able to report clear alterations in diseases associated to genetic malfunctions, specifically Huntington's disease using primary cultures [19], or in Parkinson's [7], Sanfilippo [6] and frontotemporal dementia [5] using hiPSCs.

Acknowledgements This research was supported by the European Union's Horizon 2020 research and innovation program under grant No. 964977 (project NEU-ChiP). J.S. would like to thank the Spanish Ministerio de Ciencia e Innovación under project PID2022-137713NB-C22, funded by MCIU/AEI/10.13039/501100011033 and by ERDF/EU, the Generalitat de Catalunya under grant 2021-SGR-00450, and "la Caixa" Foundation (ID 100010434) under grant agreement LCF/PR/HR19/52160007 for financial support.

References

1. Allen, N.J., Barres, B.A.: Glia—more than just brain glue. Nature **457**, 675–677 (2009). https://doi.org/10.1038/457675a
2. Ayasreh, S., Jurado, I., López-León, C., Montalà-Flaquer, M., Soriano, J.: Dynamic and functional alterations of neuronal networks in vitro upon physical damage: a proof of concept. Micromachines **13**, article 2259 (2022). https://doi.org/10.3390/mi13122259
3. Bisio, M., Bosca, A., Pasquale, V., Berdondini, L., Chiappalone, M.: Emergence of bursting activity in connected neuronal sub-populations. PLoS One **9**(9), e107400 (2014)
4. Boccaletti, S., Latora, V., Moreno, Y., Chavez, M., Hwang, D.: Complex networks: structure and dynamics. Phys. Rep. **424**, 175–308 (2006). https://doi.org/10.1016/j.physrep.2005.10.009

5. Canals, I., Comella-Bolla, A., Cepeda-Prado, E., Avaliani, N., Crowe, J.A., Oburoglu, L., Bruzelius, A., King, N., Pajares, M.A., Pérez-Sala, D., et al.: Astrocyte dysfunction and neuronal network hyperactivity in a CRISPR engineered pluripotent stem cell model of frontotemporal dementia. Brain Commun. **5**(3), fcad158 (2023)
6. Canals, I., Soriano, J., Orlandi, J.G., Torrent, R., Richaud-Patin, Y., Jiménez-Delgado, S., Merlin, S., Follenzi, A., Consiglio, A., Vilageliu, L., Grinberg, D., Raya, A.: Activity and high-order effective connectivity alterations in Sanfilippo C patient-specific neuronal networks. Stem Cell Rep. **5**(4), 546–557 (2015). https://doi.org/10.1016/j.stemcr.2015.08.016
7. Carola, G., Malagarriga, D., Calatayud, C., Pons-Espinal, M., Blasco-Agell, L., Richaud-Patin, Y., Fernandez-Carasa, I., Baruffi, V., Beltramone, S., Molina, E., Dell'Era, P., Toledo-Aral, J.J., Tolosa, E., Muotri, A.R., Garcia Ojalvo, J., Soriano, J., Raya, A., Consiglio, A.: Parkinson's disease patient-specific neuronal networks carrying the LRRK2 G2019S mutation unveil early functional alterations that predate neurodegeneration. npj Parkinson's Dis. **7**(1), 1–14 (2021). https://doi.org/10.1038/s41531-021-00198-3
8. Chiappalone, M., Vato, A., Berdondini, L., Koudelka-Hep, M., Martinoia, S.: Network dynamics and synchronous activity in cultured cortical neurons. Int. J. Neural Syst. **17**(02), 87–103 (2007)
9. Comella-Bolla, A., Orlandi, J.G., Miguez, A., Straccia, M., García-Bravo, M., Bombau, G., Galofré, M., Sanders, P., Carrere, J., Segovia, J.C., Blasi, J., Allen, N.D., Alberch, J., Soriano, J., Canals, J.M.: Human pluripotent stem cell-derived neurons are functionally mature in vitro and integrate into the mouse striatum following transplantation. Mol. Neurobiol. **57**(6), 2766–2798 (2020). https://doi.org/10.1007/s12035-020-01907-4
10. Estévez-Priego, E., Moreno-Fina, M., Monni, E., Kokaia, Z., Soriano, J., Tornero, D.: Long-term calcium imaging reveals functional development in hiPSC-derived cultures comparable to human but not rat primary cultures. Stem Cell Rep. **18**, 1–15 (2023). https://doi.org/10.1016/j.stemcr.2022.11.014
11. Estévez-Priego, E., Teller, S., Granell, C., Arenas, A., Soriano, J.: Functional strengthening through synaptic scaling upon connectivity disruption in neuronal cultures. Netw. Neurosci. **4**, 1160–1180 (2020). https://doi.org/10.1162/netn_a_00156
12. Faci-Lázaro, S., Soriano, J., Mazo, J.J., Gómez-Gardeñes, J.: Dynamical and topological conditions triggering the spontaneous activation of izhikevich neuronal networks. Chaos Solitons Fractals **172**, 113547 (2023)
13. Faci-Lázaro, S., Soriano, J., Gómez-Gardeñes, J.: Impact of targeted attack on the spontaneous activity in spatial and biologically-inspired neuronal networks. Chaos **29**, 083126 (2019). https://doi.org/10.1063/1.5099038
14. Fan, J., Meng, J., Ludescher, J., Chen, X., Ashkenazy, Y., Kurths, J., Havlin, S., Schellnhuber, H.J.: Statistical physics approaches to the complex earth system. Phys. Rep. **896**, 1–84 (2021)
15. Feinerman, O., Rotem, A., Moses, E.: Reliable neuronal logic devices from patterned hippocampal cultures. Nat. Phys. **4**(12), 967–973 (2008)
16. Feinerman, O., Segal, M., Moses, E.: Signal propagation along unidimensional neuronal networks. J. Neurophysiol. **94**(5), 3406–3416 (2005)
17. Fernández-García, S., Orlandi, J.G., García-Díaz Barriga, G., Rodríguez, M., Masana, M., Soriano, J., Alberch, J.: Deficits in coordinated neuronal activity and network topology are striatal hallmarks in Huntington's disease. BMC Biol. **18**(1), 1–16 (2020)
18. Fernández-Santiago, R., Carballo-Carbajal, I., Castellano, G., Torrent, R., Richaud, Y., Sánchez-Danés, A., Vilarrasa-Blasi, R., Sánchez-Pla, A., Mosquera, J.L., Soriano, J., et al.: Aberrant epigenome in iPSC-derived dopaminergic neurons from Parkinson's disease patients. EMBO Mol. Med. **7**(12), 1529–1546 (2015)
19. Fernández-García, S., Orlandi, J.G., Barriga, G.A.G.D., Rodríguez, M.J., Masana, M., Soriano, J., Alberch, J.: Deficits in coordinated neuronal activity and network topology are striatal hallmarks in Huntington's disease. BMC Biol. **18**, 1–16 (2020). https://doi.org/10.1186/s12915-020-00794-4
20. FitzHugh, R.: Impulses and physiological states in theoretical models of nerve membrane. Biophys. J. **1**(6), 445–466 (1961)

21. Friedrich, J., Zhou, P., Paninski, L., Staneva, V., Chklovskii, D., Pnevmatikakis, E.: Fast online deconvolution of calcium imaging data. PLOS Comput. Biol. **13**, e1005423 (2017). https://doi.org/10.1371/journal.pcbi.1005423
22. Golomb, D., Ermentrout, G.B.: Continuous and lurching traveling pulses in neuronal networks with delay and spatially decaying connectivity. Proc. Natl. Acad. Sci. **96**(23), 13480–13485 (1999)
23. Götz, J., Halliday, G., Nisbet, R.M.: Molecular pathogenesis of the tauopathies. Annu. Rev. Pathol. Mech. Dis. **14**, 239–261 (2019)
24. Grewe, B.F., Langer, D., Kasper, H., Kampa, B.M., Helmchen, F.: High-speed in vivo calcium imaging reveals neuronal network activity with near-millisecond precision. Nat. Methods **7**, 399–405 (2010). https://doi.org/10.1038/nmeth.1453
25. Grienberger, C., Konnerth, A.: Imaging calcium in neurons. Neuron **73**, 862–885 (2012). https://doi.org/10.1016/j.neuron.2012.02.011
26. Guo, T., Noble, W., Hanger, D.P.: Roles of tau protein in health and disease. Acta Neuropathol. **133**, 665–704 (2017)
27. Hernández-Navarro, L., Faci-Lázaro, S., Orlandi, J.G., Feudel, U., Gómez-Gardeñes, J., Soriano, J.: Noise-driven amplification mechanisms governing the emergence of coherent extreme events in excitable systems. Phys. Rev. Res. **3**, 1–15 (2021). https://doi.org/10.1103/physrevresearch.3.023133
28. Hernández-Navarro, L., Orlandi, J.G., Cerruti, B., Vives, E., Soriano, J.: Dominance of metric correlations in two-dimensional neuronal cultures described through a random field ising model. Phys. Rev. Lett. **118**, 1–5 (2017). https://doi.org/10.1103/PhysRevLett.118.208101
29. Idelson, M.S., Ben-Jacob, E., Hanein, Y.: Innate synchronous oscillations in freely-organized small neuronal circuits. PLoS One **5**, e14443 (2010). https://doi.org/10.1371/journal.pone.0014443
30. Izhikevich, E.M.: Simple model of spiking neurons. IEEE Trans. Neural. Netw. A publication of the IEEE Neural Networks Council **14**, 1569–1572 (2003). https://doi.org/10.1109/TNN.2003.820440
31. Izhikevich, E.M.: Which model to use for cortical spiking neurons? IEEE Trans. Neural Netw. **15**, 1063–1070 (2004). https://doi.org/10.1109/TNN.2004.832719
32. Jacobi, S., Soriano, J., Moses, E.: BDNF and NT-3 increase velocity of activity front propagation in unidimensional hippocampal cultures. J. Neurophysiol. **104**(6), 2932–2939 (2010)
33. Jacobi, S., Soriano, J., Moses, E.: BDNF and NT-3 increase velocity of activity front propagation in unidimensional hippocampal cultures. J. Neurophysiol. **104**, 2932–2939 (2010). https://doi.org/10.1152/jn.00002.2010
34. Jacobi, S., Soriano, J., Segal, M., Moses, E.: BDNF and NT-3 increase excitatory input connectivity in rat hippocampal cultures. Eur. J. Neurosci. **30**(6), 998–1010 (2009)
35. Kandel, E.R.: Dale's principle and the functional specificity of neurons. Psychopharmacol. A Rev. Prog. **1967**, 385–398 (1957)
36. Kim, M.J., Loucks, R.A., Palmer, A.L., Brown, A.C., Solomon, K.M., Marchante, A.N., Whalen, P.J.: The structural and functional connectivity of the amygdala: from normal emotion to pathological anxiety. Behav. Brain Res. **223**(2), 403–410 (2011)
37. Knöpfel, T.: Genetically encoded optical indicators for the analysis of neuronal circuits. Nat. Rev. Neurosci. **13**, 687–700 (2012). https://doi.org/10.1038/nrn3293
38. Kolagar, T.A., Farzaneh, M., Nikkar, N., Khoshnam, S.E.: Human pluripotent stem cells in neurodegenerative diseases: potentials, advances and limitations. Curr. Stem Cell Res. Ther. **15**(2), 102–110 (2020)
39. Li, M., Liu, R.R., Lü, L., Hu, M.B., Xu, S., Zhang, Y.C.: Percolation on complex networks: theory and application. Phys. Rep. **907**, 1–68 (2021)
40. Ludl, A.A., Soriano, J.: Impact of physical obstacles on the structural and effective connectivity of in silico neuronal circuits. Front. Comput. Neurosci. **14**, 1–5 (2020). https://doi.org/10.3389/fncom.2020.00077
41. Medaglia, J.D., Bassett, D.S.: Network Analyses and Nervous System Disorders. Oxford University Press (2018). https://doi.org/10.1093/acrefore/9780190264086.013.121

42. Millet, L.J., Gillette, M.U.: Over a century of neuron culture: from the hanging drop to microfluidic devices. Yale J. Biol. Med. **85**(4), 501–521 (2012)
43. Ming, Y., Abedin, M.J., Tatic-Lucic, S., Berdichevsky, Y.: Microdevice for directional axo-dendritic connectivity between micro 3d neuronal cultures. Microsyst. Nanoeng. **7**(1), 67 (2021)
44. Montalà-Flaquer, M., López-León, C.F., Tornero, D., Houben, A.M., Fardet, T., Monceau, P., Bottani, S., Soriano, J.: Rich dynamics and functional organization on topographically designed neuronal networks in vitro. iScience **25**, article 105680 (2022). https://doi.org/10.1016/j.isci.2022.105680
45. Nagumo, J., Arimoto, S., Yoshizawa, S.: An active pulse transmission line simulating nerve axon. Proc. IRE **50**(10), 2061–2070 (1962)
46. Nishikawa, S.I., Goldstein, R.A., Nierras, C.R.: The promise of human induced pluripotent stem cells for research and therapy. Nat. Rev. Mol. Cell Biol. **9**(9), 725–729 (2008). https://doi.org/10.1038/nrm2466
47. Orlandi, J.G., Soriano, J., Alvarez-Lacalle, E., Teller, S., Casademunt, J.: Noise focusing and the emergence of coherent activity in neuronal cultures. Nat. Phys. **9**(9), 582–590 (2013). https://doi.org/10.1038/nphys2686
48. O'Brien, R.J., Kamboj, S., Ehlers, M.D., Rosen, K.R., Fischbach, G.D., Huganir, R.L.: Activity-dependent modulation of synaptic AMPA receptor accumulation. Neuron **21**(5), 1067–1078 (1998)
49. Palop, J.J., Mucke, L.: Network abnormalities and interneuron dysfunction in Alzheimer disease. Nat. Rev. Neurosci. **17**(12), 777–792 (2016)
50. Paredes, R.M., Etzler, J.C., Watts, L.T., Zheng, W., Lechleiter, J.D.: Chemical calcium indicators. Methods **46**, 143–151 (2008). https://doi.org/10.1016/j.ymeth.2008.09.025
51. Renault, R., Durand, J.B., Viovy, J.L., Villard, C.: Asymmetric axonal edge guidance: a new paradigm for building oriented neuronal networks. Lab Chip **16**(12), 2188–2191 (2016)
52. Sadeh, S., Clopath, C.: Inhibitory stabilization and cortical computation. Nat. Rev. Neurosci. **22**(1), 21–37 (2021)
53. Sala-Jarque, J., Zimkowska, K., Ávila, J., Ferrer, I., Del Río, J.A.: Towards a mechanistic model of tau-mediated pathology in tauopathies: what can we learn from cell-based in vitro assays? Int. J. Mol. Sci. **23**(19), 11527 (2022)
54. de Santos-Sierra, D., Sendiña-Nadal, I., Leyva, I., Almendral, J.A., Ayali, A., Anava, S., Sánchez-Ávila, C., Boccaletti, S.: Graph-based unsupervised segmentation algorithm for cultured neuronal networks' structure characterization and modeling. Cytometry Part A **87**(6), 513–523 (2015). https://doi.org/10.1002/cyto.a.22591
55. Shein-Idelson, M., Ben-Jacob, E., Hanein, Y.: Engineered neuronal circuits: a new platform for studying the role of modular topology. Front. Neuroeng. **4**, 10 (2011)
56. Shi, M., Majumdar, D., Gao, Y., Brewer, B.M., Goodwin, C.R., McLean, J.A., Li, D., Webb, D.J.: Glia co-culture with neurons in microfluidic platforms promotes the formation and stabilization of synaptic contacts. Lab Chip **13**, 3008–3021 (2013). https://doi.org/10.1039/c3lc50249j
57. Shi, Y., Inoue, H., Wu, J.C., Yamanaka, S.: Induced pluripotent stem cell technology: a decade of progress. Nat. Rev. Drug Discov. **16**(2), 115–130 (2017). https://doi.org/10.1038/nrd.2016.245
58. Siegenfeld, A.F., Bar-Yam, Y.: An introduction to complex systems science and its applications. Complexity **2020**, 1–16 (2020)
59. Soriano, J.: Neuronal cultures: exploring biophysics, complex systems, and medicine in a dish. Biophysica **3**(1), 181–202 (2023)
60. Soriano, J., Martínez, M.R., Tlusty, T., Moses, E.: Development of input connections in neural cultures. Proc. Natl. Acad. Sci. **105**(37), 13758–13763 (2008). https://doi.org/10.1073/pnas.0707492105
61. Soriano-Paños, D., Cota, W., Ferreira, S.C., Ghoshal, G., Arenas, A., Gómez-Gardeñes, J.: Modeling communicable diseases, human mobility, and epidemics: a review. Annalen der Physik **534**(6), 2100482 (2022)

62. Stam, C.J.: Modern network science of neurological disorders. Nat. Rev. Neurosci. **15**(10), 683–695 (2014). https://doi.org/10.1038/nrn3801
63. Strogatz, S.: Nonlinear Dynamics and Chaos: With Applications to Physics, Biology, Chemistry and Engineering. CRC Press, Boca Raton, FL, USA (2018)
64. Styr, B., Slutsky, I.: Imbalance between firing homeostasis and synaptic plasticity drives early-phase Alzheimer's disease. Nat. Neurosci. **21**(4), 463–473 (2018)
65. Sukenik, N., Vinogradov, O., Weinreb, E., Segal, M., Levina, A., Moses, E.: Neuronal circuits overcome imbalance in excitation and inhibition by adjusting connection numbers. Proc. Natl. Acad. Sci. **118**(12), e2018459118 (2021)
66. Takahashi, K., Yamanaka, S.: Induction of pluripotent stem cells from mouse embryonic and adult fibroblast cultures by defined factors. Cell **126**(4), 663–676 (2006). https://doi.org/10.1016/j.cell.2006.07.024
67. Targa Dias Anastacio, H., Matosin, N., Ooi, L.: Neuronal hyperexcitability in Alzheimer's disease: what are the drivers behind this aberrant phenotype? Transl. Psychiatr. **12**(1), 257 (2022)
68. Teller, S., Estévez-Priego, E., Granell, C., Tornero, D., Andilla, J., Olarte, O.E., Loza-Alvarez, P., Arenas, A., Soriano, J.: Spontaneous functional recovery after focal damage in neuronal cultures. Eneuro **7**(1) (2020)
69. Teller, S., Estévez-Priego, E., Granell, C., Tornero, D., Andilla, J., Olarte, O.E., Loza-Alvarez, P., Arenas, A., Soriano, J.: Spontaneous functional recovery after focal damage in neuronal cultures. eNeuro **7**, article 0254–19.2019 (2020). https://doi.org/10.1523/ENEURO.0254-19.2019
70. Teller, S., Granell, C., De Domenico, M., Soriano, J., Gómez, S., Arenas, A.: Emergence of assortative mixing between clusters of cultured neurons. PLoS Comput. Biol. **10**(9), 1–17 (2014). https://doi.org/10.1371/journal.pcbi.1003796
71. Teller, S., Tahirbegi, I.B., Mir, M., Samitier, J., Soriano, J.: Magnetite-Amyloid-β deteriorates activity and functional organization in an in vitro model for Alzheimer's disease. Sci. Rep. **5**(July 2015), 1–16 (2015). https://doi.org/10.1038/srep17261
72. Teller, S., Tahirbegi, I.B., Mir, M., Samitier, J., Soriano, J.: Magnetite-amyloid-β deteriorates activity and functional organization in an in vitro model for Alzheimer's disease. Sci. Rep. **5**(1), 17261 (2015)
73. Tibau, E., Ludl, A.A., Rudiger, S., Orlandi, J.G., Soriano, J.: Neuronal spatial arrangement shapes effective connectivity traits of in vitro cortical networks. IEEE Trans. Netw. Sci. Eng. **7**, 435–448 (2020). https://doi.org/10.1109/TNSE.2018.2862919
74. Tibau, E., Valencia, M., Soriano, J.: Identification of neuronal network properties from the spectral analysis of calcium imaging signals in neuronal cultures. Front. Neur. Circ. **7**, article 199 (2013). https://doi.org/10.3389/fncir.2013.00199
75. Turrigiano, G.G., Nelson, S.B.: Homeostatic plasticity in the developing nervous system. Nat. Rev. Neurosci. **5**(2), 97–107 (2004)
76. Wang, Y., Mandelkow, E.: Tau in physiology and pathology. Nat. Rev. Neurosci. **17**(1), 22–35 (2016)
77. Yamamoto, H., Moriya, S., Ide, K., Hayakawa, T., Akima, H., Sato, S., Kubota, S., Tanii, T., Niwano, M., Teller, S., Soriano, J., Hirano-Iwata, A.: Impact of modular organization on dynamical richness in cortical networks. Sci. Adv. **4**, eaau4914 (2018). https://doi.org/10.1126/sciadv.aau4914
78. Yuste, R.: From the neuron doctrine to neural networks. Nat. Rev. Neurosci. **16**(8), 487–497 (2015). https://doi.org/10.1038/nrn3962
79. Zamora-López, G., Chen, Y., Deco, G., Kringelbach, M.L., Zhou, C.: Functional complexity emerging from anatomical constraints in the brain: the significance of network modularity and rich-clubs. Sci. Rep. **6**, 1–18 (2016). https://doi.org/10.1038/srep38424
80. Zanin, M., Papo, D., Sousa, P.A., Menasalvas, E., Nicchi, A., Kubik, E., Boccaletti, S.: Combining complex networks and data mining: why and how. Phys. Rep. **635**, 1–44 (2016)
81. Zhu, Z., Huangfu, D.: Human pluripotent stem cells: an emerging model in developmental biology. Development **140**(4), 705–717 (2013)

Cell Dynamics

Modeling and Analyzing Biological Systems Using Branching Processes

Diego Pérez-Dones, Diego Mazo-Durán, and David G. Míguez

Abstract Branching processes constitute a special type of stochastic model that are emerging as powerful tool for studying and understanding biological systems. This chapter explores the basis of the branching processes formalism, introducing the fundamental concepts of the Galton-Watson framework and its more biologically relevant extensions. We then present specific biological scenarios and examples, ranging from population growth to disease spreading and evolutionary dynamics. Finally, we provide a detailed description of our own extension of the framework to study the dynamics of organogenesis in the context of stem cell decisions. By simplifying the complexity of stem cell differentiation as a simple branching process in which cells divide via either symmetric or asymmetric types of divisions, we derive analytical solutions that enable us to quantify changes in the balance between self-renewal and terminal specification that occur during development. This chapter aims to inspire researchers to leverage the power and simplicity of branching processes as an alternative or complementary approach to studying complex biological systems.

1 Introduction

Branching processes represent a distinctive class of stochastic models that have emerged as a powerful yet straightforward mathematical framework for studying population dynamics and evolutionary processes in biology [1–5]. In contrast to more commonly used stochastic models, such as Markov Chains, Stochastic Differential Equations, and Random Walks, branching processes exhibit specific characteristics:

D. Pérez-Dones · D. Mazo-Durán · D. G. Míguez (✉)
Depto. de Física de la Materia Condensada, Instituto de Física de la Materia Condensada (IFIMAC) and Centro de Biología Molecular Severo Ochoa (CBMSO), Universidad Autónoma de Madrid, Madrid, Spain
e-mail: david.miguez@uam.es

D. Pérez-Dones
e-mail: diego.perezd@estudiante.uam.es

D. Mazo-Durán
e-mail: diego.mazo@estudiante.uam.es

© The Author(s), under exclusive license to Springer Nature Switzerland AG 2025
M. Stich and J. Carballido-Landeira (eds.), *Nonlinear Dynamics for Biological Systems*, SEMA SIMAI Springer Series 40,
https://doi.org/10.1007/978-3-031-99044-1_4

- They assume the independence of reproductive success among individuals in a population and the uniformity of this success across all individuals. This approximation significantly simplifies the mathematics, facilitating the use of probability generating functions, as later discussed in this chapter.
- Branching processes primarily focus on the average and long-term behavior of populations, addressing questions about their indefinite growth or eventual extinction. Quantities like the expected population size in each generation and the probability of extinction can be directly estimated within this formalism.
- These processes typically concentrate on populations comprising a single type of individual, in contrast to other stochastic models that may involve multiple types or complex interactions. The extension to systems with more types, such as cycling progenitors or terminally differentiated cells, will be explored in the final section of this chapter.
- Branching processes are defined in discrete generations, with each generation corresponding to a specific time step. This way, the number of individuals in each generation depends on the offspring produced by individuals in the previous generation. The transition from discrete to continuous time approximation will also be discussed in the last section.

Like other stochastic models, branching processes directly incorporate the inherent stochastic nature of key biological processes, including growth, reproduction, inheritance, and disease transmission, offering valuable insights into the behavior and characteristics of biological systems.

First introduced in the late 19th century by Francis Galton and Henry Watson, the branching processes formalism (also known as Galton-Watson processes) was initially devised to model the generational growth or disappearance of surnames in upper-class families [6]. Their early studies aimed to identify patterns or principles governing the potential extinction of surnames, considering factors such as the size of the initial population and the number of male descendants responsible for passing on the family name. The analogy between family names and biological lineages formed the basis for applying this mathematical framework to a broad range of biological phenomena.

The framework evolved from these initial studies into a more general and comprehensive theory of branching processes, finding application in various fields of biology. In population biology, it has been utilized to model growth and extinction of species [7–9], the spread of infectious diseases [10–15], and the dynamics of cell populations [16–20]. In evolutionary biology, branching processes have contributed to the understanding of speciation events, genetic diversity, and the emergence of complex traits [5, 8].

Recent advancements in computational power and statistical methods have further propelled the application of branching processes, incorporating spatial components, demographic heterogeneity and environmental factors, enabling more accurate representations of real-world biological systems.

2 Mathematical Formalism of Branching Processes

Mathematically, branching processes focus on describing the evolution of a population through discrete generations. At each generation, individuals reproduce independently, generating a random number of offspring according to a specified probability distribution. The offspring, in turn, become the parents of the next generation, and the process continues iteratively.

The system begins with an initial population N_0 of individuals, where each one has the potential to produce k descendants. The number of individuals in the n-th generation N_n (where n denotes the discrete generation index) can then be calculated based on a given probability distribution of offspring reproduction p_k. These p_k values represent the normalized probability of each individual to generate k new individuals in the next generation (i.e., the probability to generate one descendant p_1, two descendants p_2, and so on). Generally, these values remain constant across generations, implying that the reproductive behavior of individuals does not depend on time or population size—a key characteristic of Galton-Watson processes.

Next, considering these p_k constants, a function $G(z)$ is defined as the sum of all these probabilities multiplied by the corresponding powers of z (a dummy variable taking every value for which the sum converges and used merely for its powers to label distinct probabilities):

$$G(z) = p_0 \cdot z^0 + p_1 \cdot z^1 + p_2 \cdot z^2 + \cdots = \sum_{k=0}^{\infty} p_k \cdot z^k. \tag{1}$$

The function $G(z)$ is refereed as "probability generating function" (or simply PGF) and captures the probability to generate any number of descendants in a single generating function. This formulation stems from the general formula for the expectation of an arbitrary function $f(k)$ of a discrete random variable k:

$$\mathbb{E}[f(k)] = \sum_{r=0}^{\infty} f(r) \cdot P(k = r). \tag{2}$$

Then, the expectation of a discrete random variable k is defined as:

$$\mathbb{E}(k) = \sum_{r=0}^{\infty} r \cdot P(k = r). \tag{3}$$

Within branching processes framework, $f(k)$ is defined as z^k which results in generating the next identity:

$$G'(1) = \mathbb{E}(k) = \mu \tag{4}$$

Variable μ in Eq. (4) represents "mean number of offspring". This identity is easily inferred if we look at the formulation of the first derivative of Eq. (1):

$$G'(z) = \sum_{k=0}^{\infty} k \cdot z^{k-1} \cdot p_k \tag{5}$$

and evaluating for $z = 1$, we get:

$$G'(1) = \sum_{k=0}^{\infty} k \cdot p_k = \mathbb{E}(k). \tag{6}$$

This mean number of offspring (commonly refereed as μ) is a fundamental quantity that characterizes the branching process, also denoted as the "branching factor" or "fertility parameter", since it captures the long-term behavior of the system. For instance, $\mu < 1$ represents a situation where each individual generates in average less than one new individual. In this situation, the process is subcritical, and the population eventually becomes extinct. On the other hand, $\mu > 1$ represents a situation where the process is supercritical, and the population is set to grow indefinitely. If the branching factor $\mu = 1$, the process is defined as critical, and the population size fluctuates around a stable value.

Similarly to the branching factor, the variance of the offspring distribution can be directly computed from the PGF as:

$$\sigma^2 = G''(1) + G'(1) - [G'(1)]^2 \tag{7}$$

where $G'(1)$ and $G''(1)$ correspond to the first and second derivative of the PGF evaluated at at $z = 1$. It is easy to demonstrate that this calculation of the variance is equivalent to the statistical calculation:

$$\sigma^2 = \sum_{k=0}^{\infty} (k - \mu)^2 \cdot p_k. \tag{8}$$

Starting by the definition of the second derivative $G''(z)$:

$$G''(z) = \mathbb{E}[k \cdot (k-1) \cdot z^{k-2}] = \sum_{k=0}^{\infty} k \cdot (k-1) \cdot z^{k-2} \cdot p_k \tag{9}$$

and evaluating at $z = 1$ we get:

$$G''(1) = \mathbb{E}[k \cdot (k-1)] = \mathbb{E}(k^2 - k) = \mathbb{E}(k^2) - \mathbb{E}(k). \tag{10}$$

On the other hand, σ^2 is defined as the "expected values of the squared deviations from the mean", i.e.:

$$\sigma^2 = \mathbb{E}[(k - \mathbb{E}[k])^2] = \mathbb{E}(k^2) - \mathbb{E}(k)^2 \tag{11}$$

Considering Eqs. (6), (10) and (11) we can easily derive Eq. (7).

Finally, the number of individuals in the system in the n-th generation (N_n) can be expressed as dependent on the number of individuals in the system in the previous generation N_{n-1} as:

$$N_n = N_{n-1} \cdot \mu \tag{12}$$

capturing the recursive nature of the Galton-Watson process by definition, and reflecting the fact that the population size is determined by the number of individuals in the previous generation. As a consequence, based on the equation above, we can write the number of individuals at any generation as dependent on the number of individuals present two generations before:

$$N_n = N_{n-1} \cdot \mu = N_{n-2} \cdot \mu \cdot \mu = N_{n-2} \cdot \mu^2 \tag{13}$$

or three:

$$N_n = N_{n-3} \cdot \mu \cdot \mu \cdot \mu = N_{n-3} \cdot \mu^3 \tag{14}$$

and in general, as dependent on the initial number of individuals.

$$N_n = N_0 \cdot \mu^n \tag{15}$$

Following the general definition of variance and the formula for the expected variance of the number of offspring per individual (σ^2), the expected variance of the population can be calculated [21]. First, using the Law of Total Variance for randomly stopped sums we have:

$$V_n = \mu^2 \cdot V_{n-1} + \sigma^2 \cdot \mu^{n-1}. \tag{16}$$

Now taking into account that $V_1 = \sigma^2$, we can obtain V_n by substitution:

$$V_2 = \mu^2 \cdot V_1 + \sigma^2 \cdot \mu = \mu^2 \cdot \sigma^2 + \mu \cdot \sigma^2 = \mu\sigma^2 \cdot (1 + \mu) \tag{17}$$

$$V_3 = \mu^2 \cdot V_2 + \sigma^2 \cdot \mu = \mu^2\sigma^2 \cdot (1 + \mu + \mu^2) \tag{18}$$

and so on until completing the pattern, reaching:

$$
\begin{aligned}
V_n &= \mu^{n-1}\sigma^2 \cdot (1 + \mu + \mu^2 + ... + \mu^{n-1}) \\
&= \mu^{n-1}\sigma^2 \sum_{r=0}^{n-1} \mu^r \\
&= \mu^{n-1}\sigma^2 \frac{1-\mu^n}{1-\mu}
\end{aligned}
\tag{19}
$$

If $\mu = 1$, the second term in the first definition of (19) equals n, therefore ending up in the expression for the population variance as:

$$
V_n = \begin{cases} \sigma^2 n, & \text{if } \mu = 1 \\ \sigma^2 \mu^{n-1} \left(\frac{1-\mu^n}{1-\mu}\right), & \text{if } \mu \neq 1 \end{cases} \tag{20}
$$

As it happened for the expected population number, to account for an initial population different from one, we just multiply the above formulas by N_0.

A useful value commonly used in biological context is the Confidence Interval (often referred as CI), that illustrates the 95% probability of the population number falling within this interval. This can be calculated from the population variance as follows:

$$
CI = N_n \pm 1.96 \cdot \sqrt{V_n} \tag{21}
$$

2.1 Example in Population Growth

To illustrate the application of Galton-Watson processes, let's consider a simplified example of a population of beetles, where each adult beetle can produce a number of offspring with the following probabilities:

$$
p_0 = 0.2,\ p_1 = 0.5,\ p_2 = 0.3. \tag{22}
$$

A scheme of the process is shown in Fig. 1. In other words, 20% of adult beetles do not produce any offspring, 50% produce one offspring, and 30% produce two offspring. Based on these probabilities, the PGF is defined as:

$$
G(z) = p_0 \cdot z^0 + p_1 \cdot z^1 + p_2 \cdot z^2 = 0.2 + 0.5 \cdot z + 0.3 \cdot z^2. \tag{23}
$$

Based on this function, the branching factor is simply $\mu = G'(1) = (0.5 + 2 \cdot 0.3z) = 1.1$, with a variance of $\sigma^2 = G''(1) + G'(1) - [G'(1)]^2 = 0.49$. The long-term behavior of the population can be calculated based on this value, as discussed in the previous section. In this case, since $\mu > 1$, the process is supercritical and the population grows exponentially.

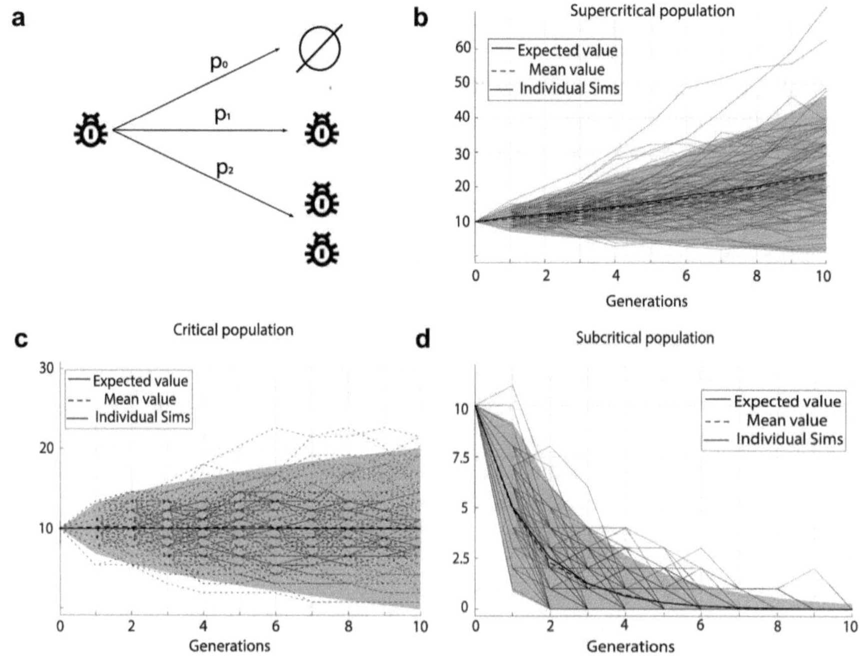

Fig. 1 Example of branching process describing population growth. **a** Schematic representation of a branching process in a beetle population. Parameters p_k represents the probabilities of having k offspring. **b–d**. Simulations of different populations with branching factor $\mu > 1$ (Supercritical), $\mu = 1$ (Critical), and $\mu < 1$ (Subcritical). Each thin dashed line represents an individual simulation, which values for each generation. Gray shade corresponds to 95% confidence interval, calculated from the population standard deviation

To examine how the size of the population evolves, let's assume an initial population size of $N_0 = 10$ beetles. For simplicity, let's only consider the first three generations and calculate the expected population size:

- Generation 1: $N_1 = N_0 G'(z) = N_0(0.5 + 2 \cdot 0.3z)$.
- Generation 2: $N_2 = N_1 G'(z) = N_0(0.5 + 2 \cdot 0.3z)^2$
- Generation 3: $N_3 = N_2 G'(z) = N_0(0.5 + 2 \cdot 0.3z)^3$

By substituting the values of $z = 1$ and $N_0 = 10$ and performing the calculations, we can determine the expected population size in each generation and the CI. A plot of how the population evolves for these values of p_k is shown in Fig. 1b.

The second example shown in Fig. 1c corresponds to a situation where the branching factor $\mu = 1$, also named as "critical", since the population fluctuates around a stable value that is equal to the initial number of individuals N_0. However, the variance now will depend on the specific combination of p_k that will result in this particular μ. For instance, in our specific case in which we only have 3 possibilities, the variance will take values between 0 (when $p_1 = 1$) and 1 (when $p_1 = 0$). This implies that, for $\mu = 1$ with chances for having any k offspring the variance will approach a maximum limit at $\max(k) - 1$.

Finally, let's input the conditions for a subcritical process. For instance:

$$p_0 = 0.6,\ p_1 = 0.3,\ p_2 = 0.1. \tag{24}$$

The PGF is now defined as:

$$G(z) = 0.6 + 0.3 \cdot z + 0.1 \cdot z^2 \tag{25}$$

and the branching factor is simply $G'(z) = (0.3 + 2 \cdot 0.1z)$ evaluated at 1, $\mu = G'(1) = 0.5$, with a variance of $\sigma^2 = G''(1) + G'(1) - [G'(1)]^2 = 0.45$. Now, on average, each individual produces fewer than one offspring in the next generation. As a result, the population size tends to decrease over time, and the process is expected to eventually become extinct. However, individual fluctuations and occasional growth are still possible, albeit less probable.

The number of individuals after each generation is:

- Generation 1: $N_1 = N_0 G'(z) = N_0(0.3 + 2 \cdot 0.1z)$.
- Generation 2: $N_2 = N_1 G'(z) = N_0(0.3 + 2 \cdot 0.1z)^2$.
- Generation 3: $N_3 = N_2 G'(z) = N_0(0.3 + 2 \cdot 0.1z)^3$.

A plot of how the population evolves is illustrated in Fig. 1d, showing that the population will disappear after certain number of generations.

Under this conditions of $\mu < 1$, we can calculate how many generations it takes for the population to become extinct. If we define the point of extinction as the generation when the number lies below 1 (which seems reasonable given that the average number of offspring per individual $\mu \leq 1$), we can establish the critical generation n_{cnt} as:

$$1 > N_0 \cdot \mu^{n_{cnt}}. \tag{26}$$

To solve this equation, we take logarithms of base μ at both sides and applying the rules of logarithms, we come to the equation:

$$n_{cnt} < -\log_{\mu}(N_0). \tag{27}$$

Using our numerical example:

$$n_{cnt} < -log_{0.5}(10) = 3.32. \tag{28}$$

So, for these values, the system will become extinct after three generations, in average.

3 Application of Branching Processes to Study Polymerase Chain Reactions

Polymerase chain reaction (PCR) is a widely used laboratory technique that allows for the amplification of specific DNA segments. It has revolutionized molecular biology, genetics, and biotechnology [22]. The analysis of PCR reactions can be effectively modeled using branching processes, providing valuable insights into the amplification dynamics and optimization of PCR protocols [23].

In a PCR reaction, a fraction of double stranded DNA serves as a template that is repeatedly replicated through a series of cycles of increase and decrease in temperature. The key enzyme involved is DNA polymerase, which synthesizes new DNA strands complementary to the template DNA. During each cycle, the DNA undergoes denaturation, annealing and extension, resulting in an exponential amplification of the target DNA segment.

The process of PCR reaction can be simplified as a direct Galton-Watson process [16], based on the following assumptions:

- each DNA molecule is identified as a parent
- a generation is identified as each cycle of the PCR
- the new generated molecules after each cycle are considered the offspring

A potential problem arises form the fact that, during the amplification, the parent molecules do not disappear after each cycle, in contrast to the typical assumption by Galton and Watson that the parental generation disappears as the new generation arises. On the other hand, since the offspring molecules are a duplicate of the parent strand, they are in principle indistinguishable. This way, we can assume that instead of surviving each cycle, parent molecules become new molecules, complying with the assumptions of Galton-Watson.

The probability distribution now is determined by the probability of successful replication and amplification at each cycle, considering the probabilities associated with DNA replication and amplification events. Consequently, the branching factor μ in this context represents the average number of DNA molecules produced in each cycle (also called average amplification rate), and its form depends on various factors such as enzyme efficiency, reaction conditions, and primer design.

These simplified models can provide quantitative predictions on the amplification efficiency, the optimal primer design and the expected number of DNA molecules at each cycle, regarding the effect of temperature, polymerase concentration and other reaction conditions [24, 25].

To illustrate how the branching framework is applied to PCR analysis, let's focus now on the specific case of a PCR reaction with perfect efficiency (meaning that a new molecule of DNA is generated per template, so the number of DNA molecules at cycle n is twice the number of DNA molecules at the preceding cycle $n-1$, i.e., $N_n = 2 \cdot N_{n-1}$). If we generalize this equality to just consider the initial number of DNA molecules (N_0), we get $N_n = 2^n \cdot N_0$.

This fixed offspring distribution of the PCR in conditions of perfect efficiency can be captured by the PGF as:

$$G(z) = \sum_{k=0}^{\infty} p_k z^k \tag{29}$$

where $p_k = \begin{cases} 0, & \text{for } k \neq 2 \\ 1, & \text{for } k = 2 \end{cases}$.

Therefore:

$$G(z) = 0 + 0 \cdot z + 1 \cdot z^2 + 0 \cdot z^3 + \cdots = z^2. \tag{30}$$

A scheme of this type of PCR reaction with perfect efficiency is shown in Fig. 2. The branching factor μ can be calculated as the first derivative of the PGF, $G'(z) = 2z$, substituting $z = 1$, we get $G'(1) = 2$. The variance of the offspring distribution can be computed using Eq. (7) as $\sigma^2 = 0$, which makes sense, since there is only one path to generate progeny, so no variability is allowed.

More complex examples in which branching processes are used to model imperfect efficiency PCRs have also been proposed. In the paper by Krawczak et al. [26], they propose a model that accounts for inefficiencies during the reaction which considers the possible replication errors which could compromise the reliability of gene diagnosis. They define probabilities p and $1 - p$ as the chances of mistaken and correct replications respectively. Being S_0 and D_0 the number of initial single-stranded or double-stranded DNA molecules, they use the branching process formalism to obtain the expected number and variance of correct replicates. In the case of the single-stranded DNA, the conceptualization is easy. The PGF is defined as follows:

$$G_S(z) = p \cdot z + (1 - p) \cdot z^2. \tag{31}$$

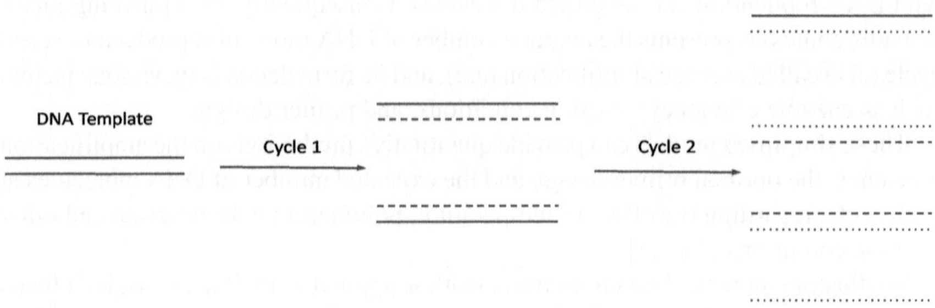

Fig. 2 Schematic representation of a PCR reaction with perfect efficiency. Starting from one DNA molecule with two strands, each strand replicates giving birth a newly synthesized DNA molecule. This type of reaction is considered to be a branching process with branching factor $\mu = 2$

This arises from the fact that, after every reaction, there is a probability p of the replication going wrong and, therefore, just obtaining one correct replicate, i.e., the original DNA, and $1 - p$ probability of a correct performance of the reaction, consequently, having two correct replicates. Following the sames steps as in the perfect efficiency reaction, we can obtain the expected number of correct replicates per individual, branching factor μ, and the offspring variance per individual, σ^2:

$$\mu = G'_S(1) = 2 - p \tag{32}$$

$$\sigma^2 = G''_S(1) + G'_S(1) - [G'_S(1)]^2 = p(1 - p). \tag{33}$$

Now for obtaining the expected population number after n cycles, we just use Eqs. (15) and (20). As they are not interested in the total number of correct DNA fragments S_n, but in the proportion of correct fragments among all replicates s_n, they propose a secondary branching process with perfect efficiency reaction, i.e., each strand gives two newly generated strands. Following the branching process rationale, $\mathbb{E}(S_n)$ and $Var(S_n)$ can be calculated using Eqs. (15) and (20) respectively. In this way, they can compute the expected proportion of correct fragments as:

$$\mathbb{E}(s_n) = \mathbb{E}(S_n/[2^n S_0]) = (1 - p/2)^n \tag{34}$$

$$Var(s_n) = Var(S_n/[2^n S_0]) \sim \frac{p(1 - p/2)^{2n-1}}{2S_0}. \tag{35}$$

In the case of the variance, the authors specify that the approximation only applies for large n. Analogously, a similar rationale for the case of double-stranded DNA can be followed.

4 Application of Branching Processes in Epidemiology

Epidemiology plays a crucial role in understanding and controlling the spread of infectious diseases within populations, and the Galton-Watson approach offer a valuable mathematical framework for modeling and analyzing the transmission dynamics of epidemic events [10–12, 14, 15]. In this context, the parents are simply identified as infected individuals, and they transmit the disease and generate new infections with certain probability. This secondary infections from each "parent" are considered the "offspring."

This way, each infected individual generates a certain number of new infections p_k determined by the disease's transmission characteristics, the population structure, and contact patterns. The variability in the number of secondary infections arises from the stochastic nature of the transmission process. This formulation allows for the estimation of key epidemiological parameters such as the basic reproduction number (R_0). R_0 is a key concept in epidemiology used to measure the intensity of disease transmission within a population, and represents the average number of secondary infections caused by a single infected individual.

Another key point is to establish the concept of "generation" in the spread of a disease (which is more of a continuum process in time). One common approach is to establish the generation time interval as the time between the onset of symptoms in an infected individual and the onset of symptoms in the individuals they infect.

One example that highlights the applications of branching processes in epidemiology focuses on the spreading of measless disease [17]. Here, the authors model the transmission dynamics of the disease in a population of susceptible individuals, formulate the PGF, the μ for the transmission, and derive the probabilities p_k based on the characteristics of transmission, such as the infectiousness of individuals and the contact rates between susceptible and infectious individuals.

Another important quantity is the average rate at which an infected individual transmits the disease to others, mathematically denoted as β. This value depends on various factors such as the mode of transmission (e.g., respiratory droplets, direct contact), the inherent infectiousness or viral load and its duration, the population density or other behaviors that facilitate close contact, and how vaccination or other of control measures like social distancing are implemented and encouraged. The value of β can be approximated from the value of μ and the p_0 (probability of no infecting another individuals) as:

$$\beta \approx \frac{\mu}{1 - p_0} \tag{36}$$

accounting for the fact that an individual can fail to transmit the disease (zero offspring), with the limitation that this approximation is only valid for diseases or conditions when p_0 is small. This way, R_0 can be written as a function of β using the following equation

$$R_0 = \beta \cdot c \cdot d, \tag{37}$$

being c the average number of contacts between infected and susceptible individuals and d being the duration of infectiousness. The dependence of the PGF on the value of β depends on the specific epidemic model being used.

Let's illustrate this with an example of a disease with the following probabilities:

$$p_0 = 0.2,\ p_1 = 0.4,\ p_2 = 0.3,\ p_3 = 0.1, \tag{38}$$

the PGF will be

$$G(z) = 0.2 + 0.4 \cdot z + 0.3 \cdot z^2 + 0.1 \cdot z^3 \tag{39}$$

and the branching factor is simply $\mu = G'(1) = (0.4 + 2 \cdot 0.3 + 3 \cdot 0.1) = 1.3$. The variance is $\sigma^2 = G''(1) + G'(1) - [G'(1)]^2 = 1.2 + 1.3 - 1.3^2 = 0.81$, while the approximated per capita rate of transmission will be:

$$\beta \approx \frac{1.3}{1-0.2} = 1.625. \tag{40}$$

Finally, it is important to take into account that, despite being equivalent in the context of Galton-Watson processes, μ and R_0 represent different quantities in the context of epidemiology. In brief, μ captures the reproductive potential of the disease at the individual level, while R_0 provides an overall measure of the potential for disease spread in a completely susceptible population, focusing more on the initial phase of an epidemic. Therefore, it would be better to define the basic reproduction number as $R_0 = \mu_{n=0}$.

5 Application of Branching Processes to Characterize the Dynamics of Stem Cell Differentiation

Developmental biology has emerged as a key research topic for the understanding of the formation and organization of developing tissues and organs, as well as the potential problems of diseases that arise when mechanisms regulating the process are defective. A cornerstone in developmental biology is the study of these regulatory processes which are in charge of stem cell maintenance and differentiation. Within this research field, cell maintenance refers to the intricate process in which cells keep their initial stem cell state, whereas cell differentiation is the process by which these stem cells respond upon intracellular and extracellular signals in order to commit to a specific fate and function (Fig. 3).

A simplistic but very powerful way to study the decision process of stem cells to self-renew or to exit the cell cycle and begin the process of differentiation is the Galton-Watson formalism. In this context, this decision is governed by distinct probabilities, which are similar for every single cell in the system. Also, the decision is taken independently by each progenitor cell regardless its previous decision and surrounding environment. Now, if progenitors are considered as parents producing two daughter cells, we can write the system as a generalized branching process where several types of progeny are possible [18, 19].

The model takes into account several cell types: progenitors (P), terminally differentiated cells (D) and cells that have died or migrated out during the process (Ψ). Now we define the probabilities of a P cell to: divide and produce two progenitor daughter ($p_2 = pp$), divide and produce one progenitor and one differentiated daughter cells ($p_1 = pd$), divide and produce two differentiated daughter cells (dd) or to undergo apoptosis or migrate out of the system (ϕ), i.e. produce 0 progenitor cells at the next generation ($p_0 = dd + \phi$).

We have also taken into account that only a percentage of progenitors are actively cycling, defined by the growth fraction γ. Given that, and assuming the conditions $pp + pd + dd = 1$ and $pp, pd, dd, \phi, \gamma \in [0, 1]$, we can define the PGF for progenitor cells $G_P(z)$ as a power series with defined probabilities for each number of offspring:

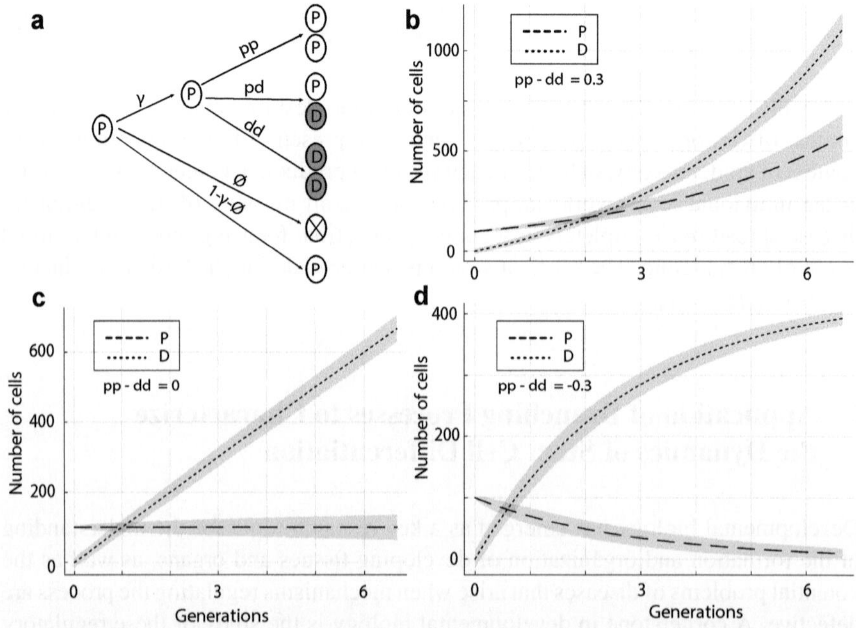

Fig. 3 Branching process applied to stem cell differentiation dynamics. **a** Schematic representation of stem cell proliferation and differentiation. White: Progenitor cells; Grey: Differentiated cells; Crossed: Dead cells. Notation above the lines represent probabilities of different outcomes. **b–d** Computation of the expected population number of progenitor and differentiated cells. Shaded areas represent 95% CI

$$G_P(z) = (\phi + \gamma \cdot dd)z^0 + (\gamma \cdot pd + 1 - \gamma - \phi)z^1 + (\gamma \cdot pp)z^2. \quad (41)$$

Again, the expected mean number of offspring per individual that are progenitors μ_P is defined by the first derivative of the PGF evaluated at 1 ($G'_P(1)$).

$$G'_P(z) = \gamma \cdot pd + 1 - \gamma - \phi + 2 \cdot pp \cdot \gamma \cdot z \quad (42)$$

$$\mu_P = G'_P(1) = \gamma \cdot pd + 1 - \gamma - \phi + 2 \cdot \gamma \cdot pp \quad (43)$$

Defining P_n as the number of progenitor cells at generation n, we have that the population number at $P_n = P_{n-1} \cdot G'_P(1) = P_{n-1} \cdot \mu_P$.

Now, applying the identity $pp + dd + pd = 1$ to the first derivative, we get:

$$\mu_P = \gamma(1 - pp - dd) + 1 - \gamma - \phi + 2 \cdot \gamma \cdot pp \quad (44)$$

$$\mu_P = 1 - \phi + \gamma(pp - dd). \quad (45)$$

Finally substituting μ_P in the equation for the number of cells:

$$P_n = P_{n-1}(1 - \phi + \gamma(pp - dd)). \tag{46}$$

After iterating, these equations for every generation we obtain the equations for the number of progenitor cells, at generation n with the form:

$$P_n = P_0(1 - \phi + \gamma(pp - dd))^n. \tag{47}$$

Then, for differentiated cells, the same procedure can be applied but with some modifications. The PGF for differentiated cells ($G_D(z)$) is computed as:

$$G_D(z) = (1 - \gamma + \gamma \cdot pp) + \gamma \cdot pd \cdot z + \gamma \cdot dd \cdot z^2. \tag{48}$$

In this case, non-cycling cells are excluded as they only contribute to the progenitor pool.

Considering that the first derivative of the probability generating function (PGF) at 1 equals the mean expected number of offspring per individual, we find that the number of newly generated differentiated cells is computed as $P_{n-1} \cdot G'_D(1)$. As stated before for progenitor cells, we can establish $G'_D(1) = \mu_D$. To account for the total number of differentiated cells at the n-th generation, we include the number of differentiated cells already in the pool D_{n-1}, resulting in $D_n = D_{n-1} + P_{n-1} \cdot \mu_D$. Using the equality to 1 of the sum of all probabilities, we simplify to:

$$D_n = D_{n-1} + P_{n-1} \cdot \mu_D = D_{n-1} + P_{n-1}(\gamma(1 + dd - pp)). \tag{49}$$

Iterating over n, we get:

$$D_n = D_0 + P_0 \cdot \mu_D(1 + \mu_P + \mu_P^2 + ... + \mu_P^{n-1}) = D_0 + P_0 \cdot \mu_D \sum_{i=0}^{n-1} \mu_P^i. \tag{50}$$

For dead cells, the PGF is calculated as:

$$G_\Psi(z) = 1 - \phi + \phi z. \tag{51}$$

Taking the first derivative yields:

$$G'_\Psi(1) = \phi = \mu_\Psi. \tag{52}$$

Subsequently:

$$\Psi_n = \Psi_{n-1} + P_{n-1} \cdot \phi. \tag{53}$$

After iteration and simplification, we reach:

$$\Psi_n = \Psi_0 + P_0 \cdot \mu_\Psi \sum_{i=0}^{n-1} \mu_P^i. \tag{54}$$

Importantly, the summation term in Eqs. (50) and (54) is equal to:

$$\sum_{i=0}^{n-1} \mu_P^i = \frac{1 - \mu_P^n}{1 - \mu_P} \tag{55}$$

and, therefore, simplifying some terms,

$$D_n = D_0 + P_0\gamma(1 + dd - pp)\frac{1 - (1 - \phi + \gamma(pp - dd))^n}{\gamma(pp - dd) - \phi} \tag{56}$$

$$\Psi_n = \Psi_0 + P_0\phi\frac{1 - (1 - \phi + \gamma(pp - dd))^n}{\gamma(pp - dd) - \phi}. \tag{57}$$

Further simplification of terms and considering Eq. (47) results in:

$$D_n = D_0 + (P_n - P_0)\frac{\gamma(1 + dd - pp)}{\gamma(pp - dd) - \phi} \tag{58}$$

$$\Psi_n = \frac{\phi(P_n - P_0)}{\gamma(pp - dd) - \phi}. \tag{59}$$

A key conclusion from the equations is their dependence on n through the number of progenitor cells, aligning with the role of progenitors as the sole contributors to offspring.

Within this branching process framework, the expected variance for progenitor cells $V_{P,n}$ is inferred directly from the formula in Sect. 2:

$$V_{P,n} = \begin{cases} P_0 \cdot \sigma_P^2 n, & \text{if } \mu_P = 1 \\ P_0 \cdot \sigma_P^2 \mu_P^{n-1}\left(\frac{1-\mu_P^n}{1-\mu_P}\right), & \text{if } \mu_P \neq 1. \end{cases} \tag{60}$$

Thus, substituting V_n for $V_{P,n}$ in Eq. (21), we obtain the 95% CI for progenitor cells.

The case for differentiated or dead cells its a little bit more complex. Starting from the general definition of the population variance in branching processes, we can see that for any generation n, the variance is computed taking into account the expected number of offspring per individual, μ_D, the offspring variance, σ_D^2, and the variance and expected population number at the previous generation. In our system we know that just the progenitor cells are able to produce offspring, therefore, the variance and expected population numbers that we are interested in, to compute the population variance of differentiated cells, are the ones of progenitor cells. This can be understood if we look at the next formula:

$$V_n = \mu^2 \cdot V_{n-1} + \mathbb{E}(N_{n-1}) \cdot \sigma^2. \tag{61}$$

This formula is derived from the Law of Total Variance of a randomly stopped sum, considering we have two random discrete variables (one for the family size, which will have the same μ and σ^2 for every individual, and the other for the population number).

For differentiated cells, the sum is stopped by the number of progenitors at the previous generation, therefore we start from the next formula

$$V_{D,n} = \mathbb{E}_{P_{n-1}}[Var(D_n|P_{n-1})] + Var_{P_{n-1}}[\mathbb{E}(D_n|P_{n-1})]. \tag{62}$$

On the left hand side we have the expectation over P_{n-1} of the variance of D_n given there are P_{n-1} individuals. This variance can be computed as $\sigma_D^2 \cdot \mathbb{E}(P_{n-1})$ according to the properties of conditional expectations and variances. On the right hand side we have the variance over P_{n-1} of the expectation of D_n given there are P_{n-1} individuals. Here again, using the same properties as before, we get $\mu_D^2 \cdot V_{P,n-1}$. Combing these two together we reach the final formula for the variance of differentiated cells:

$$V_{D,n} = \sigma_D^2 \cdot \mu_P^{n-1} P_0 + \mu_D^2 \cdot V_{P,n-1}. \tag{63}$$

Here again, substituting V_n for $V_{D,n}$ in Eq. (21), we can calculate the 95% CI.

In this context, the time between generations is directly the cell cycle duration of the progenitor cells, T. So far, we have calculated the number of cells depending on the generation number n, which is a positive integer. However, cells do not divide all at the same time in discrete steps, but they can divide at any point in time. If we consider that the beginning of the population occurs at a time $t = 0$ (when we have N_0 cells), they will start dividing continuously as t increases. In the context of a quantized branching process (as seen so far), we only measure the population every T hours (or the appropriate unit for the variable), and each measurement corresponds to a generation. That way, the timestamps at which we measure those populations are $t_q = n \cdot T$. That means that the cell cycle duration is the time that the system needs to go from generation n to generation $n + 1$.

In a branching process, where cells have a mean offspring μ after a time T, the population N at generation n is given by Eq. (15). This equation just gives values of the population at quantized times $t_q = n \cdot T$, so the equation can be rewritten as

$$N_{t_q} = N_0 \cdot \mu^{\frac{t_q}{T}} \tag{64}$$

in terms of the quantized times and the cell cycle duration. Now, we can generalize this expression so that we can calculate N at any time value t, not just $t_q = n \cdot T$. Doing so, we obtain a function that gives the population of the branching process over time

$$N(t) = N_0 \cdot \mu^{\frac{t}{T}} \tag{65}$$

where t is a real positive number indicating the time from the beginning of the experiment (at $t = 0$, when $N = N_0$) in the same units as T. In Fig. 4b we can see a plot of the Eq. (65) function showing the continuous time evolution of the population over time in a simulation.

We can see that the points given by Eq. (15) yield within the line, which leads to the conclusion that Eq. (65) correctly reproduces the branching process in the generation measurements (at $t = t_c$). The interpolation is valid if we consider a realistic case in which not all cells are born together (at $t = 0$ they have different ages) and they present some variability in the cell cycle duration. As shown in Fig. 4c, Eq. (65) describes the behavior of the simulated population perfectly, with a bit of displacement due to random variability.

Considering the previous reasoning we reach a general equation for progenitor cells at any time t:

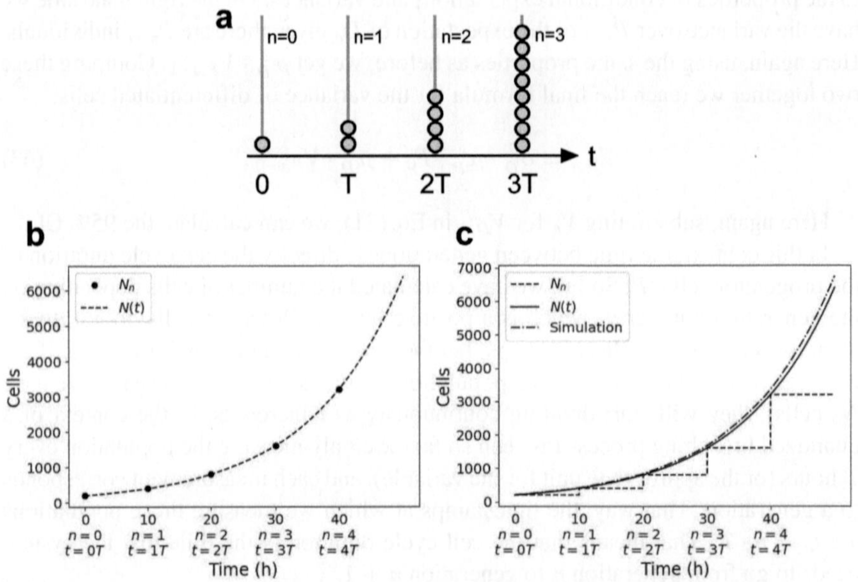

Fig. 4 Rationale behind continuous time transformation. **a** Representation of a simple branching process over time illustrating the equivalency between the generation number and the t_q timestamps. Each generation n corresponds to a specific number of cell cycles T. **b** Equivalence between discrete and continuous time branching process. Each dot represents cell population computed with discrete generations and dashed line represents cell population computed over continuous time. **c** Synchronous versus asynchronous population growth. Dashed line represents cell population computed over continuous time of a perfectly synchronized pool with all cells dividing at the same time. Dotted-dashed line represents a simulation of an asynchronous cell population with variability in cell age and dividing times. Continuous line represent cell populations computed a using branching equations for continuous time

$$P(t) = P_0(1 - \phi + \gamma(pp - dd))^{\frac{t}{T}} \tag{66}$$

Now, reversing the main goal of branching processes, we are interested in correctly calculating the average cell cycle and mode of division that, as established before, is hard to achieve using experimental techniques.

By just rewriting Eqs. (58) and (66) and solving for $pp - dd$ we obtain:

$$pp - dd = \frac{\phi \Delta D + \gamma \Delta P}{\gamma(\Delta D + \Delta P)}, \tag{67}$$

with $\Delta D = D_n - D_0$ and $\Delta P = P_n - P_0$.

For the average length of the cell cycle of the cycling progenitors we obtain:

$$T = t \cdot \frac{\log(1 - \phi + \gamma(pp - dd))}{\log \frac{P(t)}{P_0}}. \tag{68}$$

One of the most interesting features regarding the mode of division in this system is that, due to the mathematical equivalence between pp, dd, and pd, multiple combinations of values for these variables can arise for the same $pp - dd$ value. However, although these differences do not produce changes in the number of each cell type, they could represent significant variations in the system's biological self-regulation, acting in diverse ways depending on its tissue context. In other words, it may represent changes in cellular decisions based on the extracellular signals orchestrating the distribution of morphogens that determine cell fate.

On the other hand, if we consider this process as purely stochastic and independent probability for the fate of each daughter cell after mitosis, the decision between each possible fate for every single newborn cells can be defined as a binomial distribution depending on the n number if trials, with the form:

$$x \sim B\left(n, \frac{1 + pp - dd}{2}\right). \tag{69}$$

Defining each outcome as "success" or "failure", being "success" the probability of remaining proliferative and "failure" the probability of differentiate, and naming them q and $1 - q$ respectively, we can easily define q in terms of rate of division:

$$q = \frac{1 + pp - dd}{2}. \tag{70}$$

These two equations define this process as a Bernoulli trial, where the random variable x is a function of the n number of trials and the "success" probability. The probability distribution is therefore defined as:

$$P(X = x) = \binom{n}{x} \left(\frac{1 + pp - dd}{2}\right)^x \cdot \left(1 - \frac{1 + pp - dd}{2}\right)^{n-x}. \tag{71}$$

From this equation, we can obtain the probability after $n = 2$ trials of both daughter cells being either progenitor ($x = 2$) or differentiated ($x = 0$), by substituting n and x:

$$P(X=2) = \binom{2}{2}\left(\frac{1+pp-dd}{2}\right)^2 \cdot \left(1-\frac{1+pp-dd}{2}\right)^{2-2} = \left(\frac{1+pp-dd}{2}\right)^2 \quad (72)$$

$$P(X=0) = \binom{2}{0}\left(\frac{1+pp-dd}{2}\right)^0 \cdot \left(1-\frac{1+pp-dd}{2}\right)^{2-0} = \left(\frac{1+dd-pp}{2}\right)^2. \quad (73)$$

Alternatively, if after two trials just one cell remains as progenitor:

$$\begin{aligned} P(X=1) &= \binom{2}{1}\left(\frac{1+pp-dd}{2}\right)^1 \cdot \left(1-\frac{1+pp-dd}{2}\right)^{2-1} \\ &= 2\left(\frac{1+pp-dd}{2}\right) \cdot \left(1-\frac{1+pp-dd}{2}\right). \end{aligned} \quad (74)$$

This equation can be rearranged as:

$$P(X=1) = (1+pp-dd) \cdot \frac{1-pp+dd}{2} = \frac{1}{2}\left(1-(pp-dd)^2\right). \quad (75)$$

This probability can be interpreted as the the rate of pd asymmetric divisions. In a system where the previously defined conditions apply and the rate of cell death or apoptosis can be neglected, the probability can be defined in terms of change of progenitor and differentiated cells using Eq. (67):

$$pd = \frac{1}{2}\left(1-\left(\frac{\Delta P}{\Delta D+\Delta P}\right)^2\right) = \frac{\Delta D \cdot (2 \cdot \Delta P + \Delta D)}{2 \cdot (\Delta P + \Delta D)^2}. \quad (76)$$

Rearranging Eq. (67), assuming $\phi = 0$ and taking into account the equality $pp + pd + dd = 1$, we obtain:

$$pp - dd = 1 - pd - 2 \cdot dd = \frac{\Delta P}{\Delta D + \Delta P}. \quad (77)$$

Solving the previous equation for dd and substituting pd by Eq. (76) we get:

$$dd = \frac{\Delta D}{2 \cdot (\Delta D + \Delta P)} \cdot \frac{2 \cdot \Delta P + 2 \cdot \Delta D - 2 \cdot \Delta P - \Delta D}{2(\Delta D + \Delta P)} = \left(\frac{\Delta D}{2(\Delta D + \Delta P)}\right)^2. \quad (78)$$

Finally, following the same rationale, we can obtain pp by substituting Eq. (78) in Eq. (67):

$$pp = \frac{\Delta P}{\Delta D + \Delta P} + \frac{\Delta D^2}{4(\Delta D + \Delta P)^2} \tag{79}$$

and rearranging the terms:

$$pp = \frac{4\Delta P(\Delta D + \Delta P) + \Delta D}{4(\Delta D + \Delta P)^2} = \left(\frac{\Delta P + \frac{1}{2}\Delta D}{\Delta D + \Delta P}\right)^2. \tag{80}$$

6 Discussion

Galton-Watson processes provide a fundamental framework for understanding population growth and reproduction in a stochastic setting. These processes are characterized by discrete generations, where each individual in a generation independently gives rise to a random number of offspring. While the framework is a useful tool for studying population growth and extinction, there are some limitations to their use in the context of biological systems.

First, Galton-Watson processes assume that all individuals reproduce independently of each other, but there are some biological populations where this assumption does not hold. For instance, in ecosystems of insects with a complex social structure, such as ants and bees, not all individuals are able to produce progeny, and reproduction is often limited to just a small number of individuals. Similarly, in animals that live in groups, such as wolves, the reproduction dynamics depend on social features and competition between mates. Finally, in plants, reproduction is modulated by factors such as pollination, seed dispersal and competition for resources. Systems where reproduction is not independent can also be approached using more complex branching processes that include constraints specific to each particular system.

Next, Galton-Watson processes assume that the number of offspring (μ) is constant over time. This is an important limitation in the context of biological systems, where many properties, such as the cell cycle, depend on environmental conditions or other factors such as energy or food resources that are highly variable. In the context of disease spreading, the Galton-Watson approach does not assume the possibility of disease recovery or other compartments like susceptible or recovered individuals.

Finally, Galton-Watson processes assume that there is no overlapping between individuals and progeny. In other words, they assume situations where the time between generations is larger than the life expectancy of the parents or when the parent dies or disappears before the progeny is born (such as dividing cells or bacteria). In situations where this is not the case, Galton-Watson processes can still be used, taking into account a correction to add the survival rate of the parents. For instance, a type of subcritical branching process with overlapping generations in continuous time can be used to account for individuals who can produce offspring at different ages. As mentioned above, when progeny and parent are identical, such as in PCR reactions, we can assume that the parent disappears and a new individual is born, fulfilling this way the conditions for the general Galton-Watson approach [27].

Despite the mentioned limitations, Galton-Watson processes have been used very successfully in biology to study the dynamics of evolving populations. As quantification becomes increasingly important in the context of modern molecular biology, this powerful approach will likely be used more often to quantify population dynamics in cancer [28, 29], development [30, 31], and bacterial infection. A direct application of Galton-Watson processes is the study of growth and development of organoids, either to predict the number of cells that will be present in each generation, to predict the probability of survival of the whole organoid, or to study the effect of different treatments on the probability of cell survival and the number of cells produced in each generation. Finally, a very important question in biology is why organs develop as they do in terms of cell cycle and differentiation dynamics. Perhaps the Galton-Watson approach will allow us to decipher if the combination of the cell cycle and mode of division measured in a developing organ is optimal in terms of robustness, resources, or other more complex spatial or temporal variables.

References

1. Athreya, K.B., Vidyashankar, A.N.: Branching processes. In: Handbook of Statistics, pp. 35–53. Elsevier (2001). https://doi.org/10.1016/S0169-7161(01)19004-8
2. Dorman, K.S., Sinsheimer, J.S., Lange, K.: In the garden of branching processes. SIAM Rev. **46**, 202–229 (2004). https://doi.org/10.1137/S0036144502417843
3. Jagers, P., Sagitov, S.: General branching processes in discrete time as random trees. Bernoulli **14** (2008). https://doi.org/10.3150/08-BEJ138
4. Wu, X.: Branching Processes with Biological Applications. Rice University (2010)
5. Kimmel, M., Axelrod, D.E.: Branching processes in biology. In: Interdisciplinary Applied Mathematics, 2 edn. Springer, New York (2015)
6. Watson, H.W., Galton, F.: On the probability of the extinction of families. J. Anthropol. Inst. Great Britain Ireland **4**, 138 (1875). https://doi.org/10.2307/2841222
7. Aldous, D., Popovic, L.: A critical branching process model for biodiversity. Adv. Appl. Probab. **37**(4), 1094–1115 (2005)
8. Haccou, P., Jagers, P., Vatutin, V.A.: Branching processes: variation, growth, and extinction of populations, Cambridge studies in adaptive dynamics. Cambridge University Press, Cambridge, New York (2005)
9. Molina, M., Corbacho, C., Mota, M.: A stochastic model to mathematically describe the dynamics of long-lived raptor species. Biomath. Commun. Suppl. (2023)
10. Jacob, C., Viet, A.F.: Epidemiological modeling in a branching population. Math. Biosci. **182**, 93–111 (2003). https://doi.org/10.1016/S0025-5564(02)00160-8
11. Jacob, C.: Branching Processes: Their Role in Epidemiology. IJERPH **7**, 1186–1204 (2010). https://doi.org/10.3390/ijerph7031204
12. Fathi Vajargah, B., Moradi, M.: Period dependent branching process and its applications in epidemiology. Infect. Genet. Evol. **11**, 1225–1228 (2011). https://doi.org/10.1016/j.meegid.2011.03.001
13. Lashari, A.A., Trapman, P.: Branching process approach for epidemics in dynamic partnership network. J. Math. Biol. **76**, 265–294 (2018)

14. Hendy, S., Steyn, N., James, A., Plank, M.J., Hannah, K., Binny, R.N., Lustig, A.: Mathematical modelling to inform New Zealand's COVID-19 response. J. R. Soc. New Zeal. **51**, S86–S106 (2021). https://doi.org/10.1080/03036758.2021.1876111
15. Mitrofani, I.A., Koutras, V.P.: A branching process model for the novel coronavirus (Covid-19) spread in Greece. IJMO 63–69 (2021). https://doi.org/10.7763/IJMO.2021.V11.779
16. Sochivko, D.G., Fedorov, A.A., Varlamov, D.A., Kurochkin, V.E., Petrov, R.V.: Simulation of the PCR amplification as two-type-particle branching process. Dokl. Biochem. Biophys. **434**, 239–241 (2010). https://doi.org/10.1134/S1607672910050054
17. Brauer, F., Van den Driessche, P., Allen, L.J.: Mathematical epidemiology, vol. 1945, pp. 3–17. In: Wu, J. (ed.) Springer, Berlin (2008)
18. Míguez, D.G.: A branching process to characterize the dynamics of stem cell differentiation. Sci. Rep. **5**, 13265 (2015). https://doi.org/10.1038/srep13265
19. Ledesma-Terrón, M., Peralta-Cañadas, N., Míguez, D.G.: FGF2 modulates simultaneously the mode, the rate of division and the growth fraction in cultures of radial glia. Development **147**(14), dev189712 (2020)
20. Wang, R.Y., Kimmel, M.: Analysis of a countable-type branching process model for the tug-of-war cancer cell dynamics. arXiv preprint arXiv:2302.08695 (2023)
21. Allen, L.J.: Branching processes. In: Encyclopaedia of Theoretical Ecology (2012)
22. Erlich, H.A.: Polymerase chain reaction. J. Clin. Immunol. **9**, 437–447 (1989). https://doi.org/10.1007/BF00918012
23. Sun, F.: The polymerase chain reaction and branching processes. J. Comput. Biol. **2**, 63–86 (1995). https://doi.org/10.1089/cmb.1995.2.63
24. Piau, D.: Immortal branching Markov processes: averaging properties and PCR applications. Ann. Probab. **32** (2004). https://doi.org/10.1214/aop/1078415838
25. Lalam, N., Jacob, C., Jagers, P.: Modelling the PCR amplification process by a size-dependent branching process and estimation of the efficiency. Adv. Appl. Probab. **36**, 602–615 (2004). https://doi.org/10.1239/aap/1086957587
26. Krawczak, M., Reiss, J., Schmidtke, J., Rösler, U.: Polymerase chain reaction: replication errors and reliability of gene diagnosis. Nucleic Acids Res. **6**, 2197–2201 (1989). https://doi.org/10.1093/nar/17.6.2197
27. Guo, H., Zhang, M.: Limit properties of subcritical CMJ processes about the coming generation. Sci. Sin.-Math. **45**, 301–310 (2015). https://doi.org/10.1360/012015-12
28. Durrett, R., Durrett, R.: Branching Process Models of Cancer, pp. 1–63. Springer International Publishing (2015)
29. Avanzini, S., Antal, T.: Cancer recurrence times from a branching process model. PLoS Comput. Biol. **15**(11), e1007423 (2019)
30. Krapivsky, P.L., Redner, S.: Immortal branching processes. Phys. A Stat. Mech. Appl. **571**, 125853. https://doi.org/10.1016/j.physa.2021.125853
31. Genthon, A., García-García, R., Lacoste, D.: Branching processes with resetting as a model for cell division. J. Phys. A Math. Theor. **55**, 074001 (2022). https://doi.org/10.1088/1751-8121/ac491a

Stochastic Reaction-Diffusion Systems in Biophysics: Towards a Toolbox for Quantitative Model Evaluation

Gregor Pasemann, Carsten Beta, and Wilhelm Stannat

Abstract We discuss a statistical toolbox for a quantitative model evaluation of stochastic reaction-diffusion systems modeling space-time evolution of biophysical quantities on the intracellular level. Starting from space-time data $X_N(t, x)$, as, e.g., provided in fluorescence microscopy recordings, we discuss basic modelling principles for conditional mean trend and fluctuations in the class of stochastic reaction-diffusion systems, and subsequently apply statistical inference methods for parameter estimation. With a view towards application to real data, we discuss estimation errors and confidence intervals, in particular in dependence of spatial resolution of measurements, and investigate the impact of misspecified reaction terms and noise coefficients. We also briefly touch implementation issues of the statistical estimators. As a proof of concept we apply our toolbox to the statistical inference on intracellular actin concentration in the social amoeba *Dictyostelium discoideum*.

1 Introduction

1.1 A Motivating Example

Reaction-diffusion systems are one of the most popular classes of models in the life sciences [2, 3]. They arise as a natural description of spatially extended systems, where the dynamics at each location can be represented by a reaction kinetics that is often nonlinear, while neighboring locations in the system are coupled by diffusive

G. Pasemann (✉)
Institut für Mathematik, Humboldt Universität zu Berlin, Berlin, Germany
e-mail: gregor.pasemann@hu-berlin.de

C. Beta
Institut für Physik und Astronomie, Universität Potsdam, Potsdam, Germany
e-mail: beta@uni-potsdam.de

W. Stannat
Institut für Mathematik, TU Berlin, Berlin, Germany
e-mail: stannat@math.tu-berlin.de

© The Author(s), under exclusive license to Springer Nature Switzerland AG 2025
M. Stich and J. Carballido-Landeira (eds.), *Nonlinear Dynamics for Biological Systems*, SEMA SIMAI Springer Series 40,
https://doi.org/10.1007/978-3-031-99044-1_5

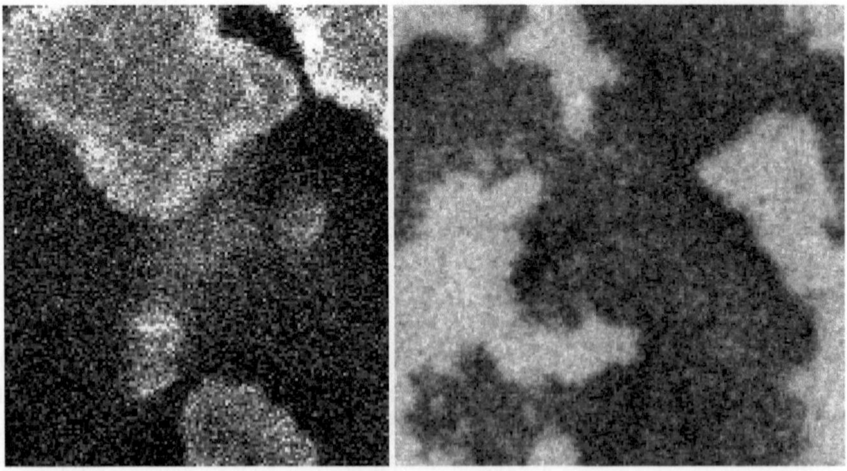

Fig. 1 Actin marker concentration in *D. discoideum* giant cells: Experimental data (left) and simulation using a FitzHugh–Nagumo type model proposed by Sergio Alonso [1] (right). Experimental data provided by Sven Flemming

transport of one or several of the reacting species. Such models are used to approximate the space-time dynamics of diverse systems and on vastly different length scales, including models of vegetation patterns [4], animal skin and tissue [5], as well as subcellular dynamical structures [6]. As most of these systems are inherently noisy, such models are often extended by including noise terms into the reaction kinetics, thus turning them into stochastic reaction-diffusion systems (Fig. 1). Here, we will focus, as a motivating example, on intracellular wave patterns that emerge in many different cell types and have been associated with essential cellular functions, such as motility, phagocytosis, and cell division [7]. To date, it is not clear, whether the same generic mechanism drives the formation of waves in different cell types, or whether fundamentally different model structures are required to address the observations across different species. Also on the molecular level, different players were found to be essential for wave formation in different cell type. Nevertheless, in most cases, the protein actin, which serves as the building block of one of the major classes of cytoskeletal filaments and is highly conserved across all eukaryotic cell types, is associated with wave formation, so that these waves are commonly denoted as "actin waves" [8, 9].

A widely used model organism to study the dynamics of actin waves is the social amoeba *Dictyostelium discoideum* (*D. discoideum*) [10]. Since the first observations of actin waves in *D. discoideum* in the 1990s, their dynamics and molecular composition have been studied in great detail [11–13]. In the wake of this experimental progress, many models have been proposed to describe the dynamics of actin waves in *D. discoideum*. These models are typically of reaction-diffusion type, with some of them also including noise terms. However, their level of complexity may vary widely, mostly with respect to the number of signaling components included in the reaction terms. The more biologically oriented models range from 4 or 5 components up to dozens of signaling molecules. In contrast, also simplified models have been proposed with the aim of capturing the essential dynamical features of the intracellular wave phenomena with the minimal set of mathematical equations. These models are often inspired by generic pattern forming reaction-diffusion systems, such as the FitzHugh-Nagumo, Gray-Scott, or Brusselator models [9].

While numerical simulations of many of these models show close agreement with experimental recordings of actin waves, it remains an open question, which of the models performs best in the light of available data. How robust is the agreement with experimental data? Are the model parameters realistic? Which are the essential nonlinearities and coupling terms that need to be included in the model? Developing mathematically rigorous approaches to infer parameter values as well as the structure of reaction terms and noise contributions in stochastic reaction-diffusion models from experimental data will greatly contribute to a more systematic model design to advance our understanding of intracellular actin waves and their associated biological functions.

As a prototypical example of quantitative model evaluation in a stochastic reaction-diffusion system modelling the space-time evolution of intracellular waves, we will consider here a spatially extended stochastic FitzHugh-Nagumo-type model of actin wave formation in *D. discoideum* [14]. The model was initially introduced to describe wave-induced cytofission events in *D. discoideum* [1] and was developed by extending a model of amoeboid motility based on a noisy bistable switch [15, 16]. In order to assess the quality of this model, beyond a visual comparison of simulation output with measurements, a crucial step is to estimate the unknown model parameters together with corresponding confidence intervals. To this end, a mathematical theory for parameter estimation of stochastic reaction-diffusion systems has been developed in [14] and was applied primarily to the estimation of the effective diffusivity of the actin concentration/density and the estimation of other reaction constants.

The data in this example was given in terms of confocal laser scanning microscopy recordings of giant *D. discoideum* cells generated by electric pulse-induced fusion of individual cells [17]. The use of giant cells provides a setting, where wave dynamics can be recorded over extended cortical areas without having to take boundary effects into account and where system size effects can be systematically explored [18, 19]. Quantitative parameter estimates like this provide first opportunities to perform sanity checks on a proposed model by comparing the obtained numbers with know literature values or physical limits in their order of magnitude. We will turn to the results of the statistical analysis of this example in more detail in Sect. 6 below.

1.2 Context and Outline

Most of the existing literature on the application of reaction-diffusion systems deals with modelling, numerical approximation and (statistical) analysis of simulation output [20, 21]. In fact, identifying and validating differential equations explaining the space-time evolution of physical observables is a classical problem in the natural and the engineering sciences and there is a huge body of literature devoted to data-driven approaches, see in particular [22] and the references therein. When it comes to the explanation of space-time data obtained in real experiments however, there is little existing literature on how to perform a quantitative validation of these models. This is partly due to the fact that the statistical inference of stochastic reaction-diffusion systems, and more general stochastic partial differential equations (SPDEs), is far from being well-developed. Another reason is that space-time data of sufficiently fine space time resolution had not been available or not so easily tractable in the past. Notably, there are early examples [23, 24] of SPDE-based data analysis in the field of groundwater hydrology by means of methods that are very close to ours. More recently, a cell repolarization model was considered in [25]. We believe that a theoretical foundation of the statistical methods in use can improve the understanding of phenomena observed in the data, while the interplay between experiments, simulations and theoretical analysis may also lead to fruitful new directions within mathematical research.

It is the main purpose of this work to put the approach persued in [14] into the larger perspective of developing statistical tools for the quantitative model evaluation of stochastic-reaction diffusion systems in biophysics. We gather general principles for the dynamical and statistical modelling of such systems, and we provide an accessible overview of theoretical guarantees for the resulting models present in the mathematical literature, both well-established and recent, and explain how they can be applied.

We structure this work as follows: In Sect. 2 we discuss how the dynamics of a random field $X(t, x)$ is modelled as a stochastic partial differential equation, and discuss different commonly used terms for the drift and the noise. In Sect. 3, we provide a method to estimate unknown parameters in such a model. Section 4 is devoted to structural properties of the resulting estimators that can be proven with mathematical rigour. In particular, we discuss the principal differences between diffusion and reaction terms, the high resolution limit versus the large time limit, and the impact of model misspecification. In Sect. 5, we comment on the time discretization of some integrals appearing in the estimation method. Finally, Sect. 6 shows an application of our method to *D. discoideum* giant cell data.

2 The Dynamical Model—Stochastic Reaction–Diffusion Equations

2.1 *General Considerations for Space-Time Random Fields*

We start with discussing general principles for the derivation and further assessment of stochastic evolution equations

$$\partial_t X = \mathcal{F}(X) + \xi \quad (1)$$

as mathematical models describing space-time evolution of a given physical observable. Here, X is thought of as a random field in space time $X(t, x)$, and the basic decomposition[1]

$$\mathcal{F}(X)(t, x) \approx \frac{1}{h}\mathbb{E}\left[X(t+h, x) - X(t, x) \mid X(s, \cdot), s \leq t\right], \quad (2)$$

describing the mean trend in the time evolution of X, given the information of the past, $X(s, \cdot)$, $s \leq t$, and

$$\xi(t, x) \approx \frac{1}{h}\left(X(t+h, x) - X(t, x) - \mathcal{F}(X)(t, x)h\right) \quad (3)$$

models the corresponding remainder as fluctuations in terms of a stochastic process $\xi(t, x)$. Besides theoretical considerations on the overall structure of the drift term $\mathcal{F}$ and the noise term ξ on the r.h.s. of (1), our aim here is to provide additional statistical tools for a quantitative model evaluation of (1), given space time data $X_N(t, x)$ whose time-evolution should be explained by (1).

In principle, a first expansion of $\mathcal{F}(X)$ can be achieved separating linear from nonlinear components. A simplifying assumption is that the linear part dominates the nonlinear part. What "dominating" means will be made precise in Sect. 4.4. In this work, we restrict to the widely used case $\mathcal{F}(X) = \vartheta_0 \Delta X + F_\vartheta(X)$, i.e. spatial interaction within the drift is governed by an isotropic diffusion term of intensity ϑ_0, and $F_\vartheta(X)$ contains lower order (e.g. reaction) terms present in the drift, as explained in the next section. A common approach to obtain the noise term $\xi(t, x)$ which models the (conditional) fluctuations is to match the covariance of $\frac{1}{h}\left(X(t+h, x) - X(t, x) - \mathcal{F}(X)(t, x)h\right)$ to that of a Gaussian process (additive noise), or to additionally weight such a process with X-dependent coefficients (multiplicative noise).

2.2 *Stochastic Reaction-Diffusion Models*

Based on these considerations, let us next fix the precise mathematical setting for our analysis. Our quantity of interest is a random field $X(t, x)$, where $t \in [0, T]$ and $x \in \mathcal{D}$ for some domain $\mathcal{D} \subset \mathbb{R}^d$. We restrict to the case that $\mathcal{D} = [0, L_1] \times \cdots \times [0, L_d]$ is a d-dimensional rectangle with periodic boundary conditions. We assume that the temporal evolution of X is governed by three components:

$$\partial_t X = \vartheta_0 \Delta X + F_\vartheta(X) + \xi \quad (4)$$

[1] Here and in the sequel, $\mathbb{E}[X]$ denotes the expected value of a random element X.

- $\vartheta_0 \Delta X$ describes the diffusive coupling, $\Delta = \sum_{i=1}^{d} \partial_{x_i}^2$ being the Laplacian operator, and $\vartheta_0 > 0$ the diffusivity of the process. We highlight this term for two reasons: First, it seems to be a common feature of most (S)PDE models that are used in practice. On the other hand, it plays a distinguished role when it comes to statistical inference, as explained in Sect. 4.
- $F_\vartheta(X)$ includes all other explicitly modelled dynamical forces acting on the temporal evolution of X. Examples are given below. $\vartheta \in \mathbb{R}^p$ is a vector of parameters on which F may depend (e.g. reaction rates).
- $\xi = \xi(t, x)$ denotes random fluctuations, which may be either an inherent property of the dynamics, or may model an approximation of unresolved fast-scale components. We do not impose assumptions on ξ apart from statistical ones: In the simplest case, ξ is just space-time white noise, but additional assumptions on the spatial or temporal correlation structure can be made.

2.3 *Models for the Drift*

Example 1 (*reaction-advection-diffusion equation*) Consider

$$\partial_t X = \vartheta_0 \Delta X - \nabla \cdot (Xv) + f(X) + \xi, \tag{5}$$

where $v(t, x)$ is a transport velocity field and $f : \mathbb{R} \to \mathbb{R}$ a function describing local sources and sinks. Write $F_{\text{adv}}(X) = -\nabla \cdot (Xv)$ for the advection term and $F_{\text{reac}}(X) = f(X)$ for the source term, then this is a special case of (4) with $F = F_{\text{adv}} + F_{\text{reac}}$.

Example 2 (*activator-inhibitor system*) Consider the system of FitzHugh–Nagumo type [1, 14]

$$\partial_t U = D_U \Delta U + k_1 U(u_0 - U)(U - u_0 a) - k_2 V + \xi, \tag{6}$$

$$\partial_t V = D_V \Delta V + \varepsilon(bU - V), \tag{7}$$

with $U(0, x) = V(0, x) = 0$ for simplicity. Here, U is the activator, and V is the inhibitor. Later, we will treat the latter as an unobserved latent component, and consider $X = U$ as the observed random field. This model can be used to generate various dynamical patterns, including traveling waves, bistability and excitability. $D_U, D_V > 0$ are the diffusivity constants of U and V, whereas $k_1, k_2, \varepsilon, b > 0$ denote reaction rates, and $u_0 > 0$ and $0 < a < 1$ determine the stable and unstable fixed points of the reaction model. This system can be cast into the form (4): Note that by the variation of constants formula, V admits an explicit representation

$$V(t, x) = \varepsilon b \int_0^t \int_{\mathcal{D}} e^{-\varepsilon(t-s)} G(D_V(t - s), x - y) U(s, y) \mathrm{d}y \mathrm{d}s, \tag{8}$$

where $G(t, x)$ is the Green's function for the heat equation on $\mathcal{D}$. In particular, V is determined by the activator history up to time t: We write $V = F_{\text{inh}}(U)$. Additionally, let $F_{\text{act}}(U) = f_{\text{act}}(U) = k_1 U(u_0 - U)(U - u_0 a)$. Then the activator component U satisfies an equation of the form (4) with $F = F_{\text{act}} + F_{\text{inh}}$.

Example 3 (*spatiotemporal coupling term*) Extending the previous example, we can consider moving average-type interactions of the form

$$F(X)(t, x) = \int_0^t \int_{\mathcal{D}} K(t - s, x - y)X(s, y)\mathrm{d}y\mathrm{d}s, \tag{9}$$

where $K(t, x)$ is some kernel function. If $K(t, x) = e^{-\varepsilon t} G(D_V t, x)$ with the Green's function G for the heat equation on $\mathcal{D}$, then we recover the integral term in (8).

Example 4 (*further drift models*) There is a large number of frequently used dynamical models that we did not discuss so far. If the velocity field of an advection term is further specified to be of the form $v(t, x) = -\int_{\mathcal{D}} \nabla\phi(x - y)X(t, y)\mathrm{d}y$ with some interaction potential ϕ, the resulting so-called aggregation term can describe non-local attraction and repulsion forces caused by regions of high concentration of X. Diffusion-aggregation models arise as continuum limits of interacting particle systems, describing time evolution of systems at microscopic levels, and may exhibit intricate dynamical features such as blow-up in finite time, see [26] for a discussion. Non-Markovian temporal interaction may also be modelled by delay terms.

2.4 Models for the Fluctuation

Finally, we discuss different models for the noise ξ:

Example 5 (*space-time white noise*) If there is no spatial or temporal correlation in the noise, the law of ξ is determined by $\mathbb{E}[\xi(t, x)\xi(s, y)] = \sigma^2\delta(t - s)\delta(x - y)$, where $\sigma > 0$ denotes the noise intensity.

Example 6 (*Ornstein–Uhlenbeck noise, first approach*) Here, temporal correlation is introduced by modelling ξ as a stochastic process in time:

$$\partial_t \xi = -\mu\xi + \Xi \tag{10}$$

without loss of generality with $\xi(0, x) = 0$, where $\Xi(t, x)$ is a space-time white noise, and $\mu > 0$. In this model, ξ is white in space and has an exponentially decaying correlation in time.

Example 7 (*Ornstein–Uhlenbeck noise, second approach*) Alternatively, temporally correlated noise ξ may be implicitly defined by letting $W(t, x) = \int_0^t \xi(s, x)\mathrm{d}s$ instead of $\xi(t, x)$ satisfy the Ornstein–Uhlenbeck equation (10). Compared to Example 6, the solution field X then exhibits different qualitative features. For example, X will not be differentiable in time (it is "rougher").

Example 8 (*Spatially correlated noise*) We can include spatial correlation by taking spatial averages:

$$\xi(t, x) = \int_{\mathcal{D}} b(x, y)\Xi(t, y)\mathrm{d}y \tag{11}$$

with a space-time white noise Ξ and a kernel b. A typical choice is given by

$$b_{-\gamma}(x, y) = \sum_{k=1}^{\infty} \lambda_k^{-\gamma} \Phi_k(x)\Phi_k(y), \tag{12}$$

where the Φ_k are the eigenfunctions of the Laplacian with eigenvalue λ_k on the (bounded) domain $\mathcal{D}$, and $\gamma > 0$ is a tuning parameter that governs the correlation strength. Furthermore, γ is related to the smoothness of the solution field $X(t, x)$ for fixed t, with larger values of γ leading to smoother solutions.

3 Statistical Model—Inference on Model Parameters

3.1 Parametrization of the Reaction Term

Our standing assumption is that F is known up to a finite-dimensional parameter vector ϑ. Consider the case that F depends linearly on its unknown parameters:

$$F_\vartheta(X) = \vartheta_1 F_1(X) + \vartheta_2 F_2(X) + \cdots + \vartheta_p F_p(X) + F_*(X), \tag{13}$$

where the F_i form a dictionary of components of the reaction term with unknown intensity, and F_* represents any "fixed" term with known intensity (it may be zero). The dictionary terms may arise from some truncated basis expansion, or encode prior knowledge on the dynamics (such as terms generating bistable behaviour, or even possible non-local or non-Markovian terms).

Note that the linearity in (13) refers to the parametrization of F rather than the dynamical behaviour of X. In this regard, it can be considered a convention. There is no restriction on the type of dynamical systems to which this approach can be applied. In principle, it is also possible to consider nonlinear parametrizations of F (this may be a good idea if the main interest is not in generic properties of F, but into some particular and physically interpretable parameter on which F depends nonlinearly), but the advantage in using a linear parametrization lies in both algorithmic simplicity of the resulting estimator (see Sect. 3.3), and a huge existing body of literature that provides detailed understanding and possible extensions for linear statistical models (see Sect. 3.5 below for an example).

We highlight that although the dynamics of X may be highly *nonlinear*, as explained in the previous section, assumption (13) will lead to a *linear* statistical model.

Example 9 (*reaction-advection-diffusion equation—continued*) In the setting of Example 1, the velocity field v and the reaction function f may be expanded as (or approximated with) a linear combination of known basis functions with unknown coefficients, e.g. $f(X) = \sum_{k=1}^{p} \vartheta_k f_k(X)$. If the f_k have a physical meaning, the ϑ_k have a natural interpretation of reaction rates of different forcing terms.

Example 10 (*activator-inhibitor system—continued*) In the setting of Example 2, one may argue that the level u_0 of the stable phase should be treated as known. In this case, the general scheme (13) reduces to

$$\begin{aligned} \vartheta_1 &= k_1 u_0 a, & F_1(U) &= -U(u_0 - U), \\ \vartheta_2 &= k_1, & F_2(U) &= U^2(u_0 - U), \\ \vartheta_3 &= k_2 \varepsilon b, & F_3(U)(t, x) &= -\int_0^t e^{(t-s)(D_V \Delta - \varepsilon \mathrm{Id})} U(s, x) \mathrm{d}s. \end{aligned} \tag{14}$$

On the other hand, if u_0 is treated as unknown, we can still expand F_{act} in powers of U, i.e. we write $F_{\mathrm{act}}(U) = \vartheta_1 U + \vartheta_2 U^2 + \vartheta_3 U^3$ (where the ϑ_k are different as in (14)) and estimate these coefficients together with the intensity of $F_{\mathrm{inh}}(U)$. However, treating the inhibitor parameters ε, D_V as unknown, we would leave the class of linearly parametrized statistical models determined by (13). An alternative approach is to consider the moving average representation (9) instead and approximate the kernel K therein by a linear combination of basis functions with unknown coefficients.

3.2 Observing an Approximation to X

We want to infer the unknown parameters (including the diffusivity) $\vartheta_0, \vartheta_1, \ldots, \vartheta_p$ from one realization of $X(t, x)$ for $0 \leq t \leq T$ and $x \in \mathcal{D}$. In fact, we will assume that we have only some approximation $X_N(t, x)$ of $X(t, x)$ at hand. This is a more realistic assumption than having all of X available. There are different ways to model the relation between X_N and X. Common approaches are:

- **Spectral approach**: The first N eigenmodes of X are observed, i.e.

$$X_N(t, x) = \sum_{k=1}^{N} \left(\int_{\mathcal{D}} X(t, y) \Phi_k(y) \mathrm{d}y \right) \Phi_k(x), \tag{15}$$

where the Φ_k are the eigenfunctions of the Laplacian Δ on $\mathcal{D}$ [27, 28]

- **Local approach**: One (or more) local averages of X are observed, i.e. X is tested against a point-spread function located at $x_0 \in \mathcal{D}$: In this case

$$X_N(t) = \int_{\mathcal{D}} X(t, y) h^{-1} K(h^{-1}(y - x_0)) \mathrm{d}y, \tag{16}$$

  where $K(x)$ denotes the point-spread function and $h = 1/N$ is a bandwidth parameter, modelling high precision as $N \to \infty$ [29]. Note that in the case of a single local observation concentrated at $x_0 \in \mathcal{D}$, the process $X_N(t, x) = X_N(t)$ does not even depend on $x \in \mathcal{D}$. Multiple local measurements can in principle be combined with some interpolation method to fit into the present framework (see also [30]).
- **Discrete approach**: X_N is the interpolation of evaluations of X on some discrete grid in space. An estimator of least squares type as derived below has been studied in [31, Chap. 4]. For different approaches for discretely sampled data see e.g. [32–37].

Apart from X_N alone, in order to exploit the dynamics of (4) we also need suitable approximations of the drift terms appearing on the right-hand side of that equation,[2] so that we can write the dynamics of X_N (up to a possible approximation error) as

$$\partial_t X_N = \vartheta_0 (\Delta X)_N + \vartheta_1 F_1(X)_N + \cdots + \vartheta_p F_p(X)_N + F_*(X)_N + \xi_N. \tag{17}$$

For example, for any reaction term given by a function $F_i = f_i : \mathbb{R} \to \mathbb{R}$, it may be reasonable to set $F_i(X)_N = F_i(X_N)$, i.e. to evaluate that function on the approximate process. However, terms involving differential operators should be handled with more care, because one needs to approximate derivatives here, which may lead to numerical challenges. In particular, this applies to the diffusion term.[3]

3.3 The Least Squares Method

Under the assumption of linear parametrization, the problem of estimating the unknown parameters in (17) has apparent structural similarities to classical linear regression, $\partial_t X_N - F_*(X)_N$ being the response variable, and $(\Delta X)_N, F_1(X)_N, \ldots, F_p(X)_N$ being the predictors. If the dynamic noise ξ is space-time white noise, this suggests a formal least squares method

---

[2] In fact, for constructing the estimator, we only need the terms appearing in (17) instead of the approximate process X_N itself.

[3] We note that within the spectral approach, it holds $(\Delta X)_N = \Delta(X_N)$ by design, such that this problem is avoided there.

$$\int_0^T \int_{\mathcal{D}} \left(\partial_t X_N - \vartheta_0 (\Delta X)_N - \sum_{i=1}^{p} \vartheta_i F_i(X)_N - F_*(X)_N \right)^2 \mathrm{d}x\mathrm{d}t \quad \to \quad \min, \tag{18}$$

and the resulting estimator $\hat{\vartheta}_N = (\hat{\vartheta}_0^N, \ldots, \hat{\vartheta}_p^N)^T$ of $\vartheta = (\vartheta_0, \ldots, \vartheta_p)^T$ is characterized by the linear system

$$A_N \hat{\vartheta}_N = b_N, \tag{19}$$

with components given by

$$A_{N,ij} = \int_0^T \int_{\mathcal{D}} F_i(X)_N F_j(X)_N \mathrm{d}x\mathrm{d}t, \tag{20}$$

$$b_{N,i} = \int_0^T \int_{\mathcal{D}} (\partial_t X_N - F_*(X)_N) F_i(X)_N \mathrm{d}x\mathrm{d}t, \tag{21}$$

where we write $F_0(X) = \Delta X$ in order to unify notation. This implicitly assumes that the linear model (13) does not contain redundant basis functions, i.e. the matrix A_N is invertible. In this case, the estimator $\hat{\vartheta}_N$ is uniquely defined. An important special case is the "plain diffusivity estimator"

$$\hat{\vartheta}_0^N = \frac{\int_0^T \int_{\mathcal{D}} (\Delta X)_N(t, x) \partial_t X_N(t, x) \mathrm{d}x\mathrm{d}t}{\int_0^T \int_{\mathcal{D}} (\Delta X)_N(t, x)^2 \mathrm{d}x\mathrm{d}t}, \tag{22}$$

which is obtained by assuming $F = 0$ in (4).

Let us finally discuss a general phenomenon in the derivation of our estimator that is particular to the least squares method for stochastic differential equations. As it turns out, the term $\int_0^T \int_{\mathcal{D}} (\partial_t X_N)^2 \mathrm{d}x\mathrm{d}t$ is not a finite quantity. Consequently, one may argue that the optimization problem stated in (18) is ill-posed, because the functional is just an infinite constant. However, in this case there is an easy renormalization procedure: Expanding the square in the functional stated there, we may just drop the "forbidden" term since it does not depend on any unknown parameter.[4] (Note that all other terms can be given a rigorous meaning in the sense of Itô or Lebesgue integration.) This procedure does not alter the resulting normal equations (19) or the estimator, which is well-defined even if the functional in (18) is not. For that reason, we still call $\hat{\vartheta}_N$ a least squares estimator. As long as (19) is used to calculate the estimator this does not make a difference, but for the extension of the method described in Sect. 3.5, it is implicitly understood that the least squares functional is renormalized as described.

---

[4] In the mathematical literature, it is common to derive the estimator from a maximum likelihood approach using Girsanovs theorem (as stated e.g. in [38, Sect. 7.6.4]). In this case, the renormalization is implicit.

3.4 Correlated Noise

If the noise ξ is correlated, a weighted least squares method should be used instead. For example, in the spatial correlation model (12), this means that a power of the Laplacian operator $(-\Delta)^\gamma$ should be applied to all terms within the bracket of (18). In this case, the plain diffusivity estimator (22) will be substituted by

$$\hat{\vartheta}_0^N = -\frac{\int_0^T \int_{\mathcal{D}} ((-\Delta)^{1+\gamma} X)_N(t,x)((-\Delta)^\gamma \partial_t X)_N(t,x)\mathrm{d}x\mathrm{d}t}{\int_0^T \int_{\mathcal{D}} ((-\Delta)^{1+\gamma} X)_N(t,x)^2 \mathrm{d}x\mathrm{d}t} \tag{23}$$

in that setting. Note that for $\gamma \notin \mathbb{N}$, the action of $(-\Delta)^\gamma$ is defined as the convolution with the kernel $b_\gamma(x,y)$, which is given as in (12).

3.5 A Possible Extension for Model Selection

While least squares type estimators are natural and enjoy good theoretical properties, there are many situations where it may not be a good idea to use them in plain form. As an example, we discuss the problem of model selection with a high-dimensional parameter space.

Consider the situation that we observe some phenomenon in the data (such as bistability or traveling waves), but there are different ways to write down a reaction model that generates such patterns (using e.g. different polynomials, piecewise linear functions or linear combinations of other basis functions such as trigonometric polynomials). If we consider a dictionary of various such possible functions for our reaction term F, we would expect the vector of coefficients to be *sparse*, i.e. only few entries should be different from zero. Therefore, apart from only estimating the parameters, it is of independent interest to identify the subset of non-zero coefficients in order to decide for the best reaction term.

In the classical statistical learning setup, such a problem can be tackled with the LASSO [39], which is implicitly defined by adding a regularizing term of the form $\lambda \sum_{i=1}^p |\vartheta_i|$ to the quadratic functional in (18). Here, $\lambda > 0$ is a hyperparameter that governs the model complexity. A general discussion can be found in [40]. Other approaches are possible. For example, in [22], an iteratively thresholded least squares estimator is used to account for sparsity in a related setting.

> As in classical statistical learning, the least squares estimator may serve as a baseline for more complex algorithms such as the LASSO, which can be used for model selection.

However, up to now, such methods have not been analyzed from a mathematical perspective in the context of stochastic partial differential equations.

4 Properties of the Statistical Model

In the previous section, we derived estimators for the unknown model parameters from the realization of an approximate process $X_N(t, x)$. In a next step, we will discuss how the precision of the estimators improves as N grows, i.e. as the resolution gets finer. Here we work mainly within the above mentioned *spectral approach*. There is much literature on this approach (see [41] and references therein), given that it is conceptually and technically rather simple compared to other approaches that are available, but the main aspects transfer also to the case of local [29, 30] and discrete [35] observations. Recall that in this setting, we assume to observe only N spatial eigenmodes of X as in (15) sampled continuously in time, i.e. there is no discretization error in time. This is a reasonable approximation for high-frequency data (or long time series, if the characteristic time scale for the phenomena described by the SPDE is large). Also, unless otherwise stated, we assume a noise model as in (12).

The mathematical theory of statistics for SPDEs has become a vivid field in the last years (see e.g. the survey [42] and references therein). By now, there is some literature on quantitative evaluation of statistical methods for nonlinear SPDEs, although this field is far from being saturated. We highlight [43] for an early attempt to find a general framework, [44, 45] for a study of the ergodic case, [46] on the Navier–Stokes equation as well as the more recent works [36, 47]. Stability properties of diffusivity estimation under misspecification as exposed in Sect. 4.4 below have been systematically studied in [14, 31, 48, 49].

4.1 Estimation Error Versus Spatial Resolution

As it turns out, reaction and diffusion parameter behave very differently as N grows, and this behaviour is even dimension-dependent. To illustrate this point, let us consider a model problem with a linear source term ($p = 1$ and $F_1(X) = X$ in (13), i.e. $F_\vartheta(X) = \vartheta_1 X$), such that the resulting SPDE is, in fact, linear. This model has been studied in [27, 28, 50]. The diffusivity estimator is consistent in the sense that

$$\hat{\vartheta}_0^N = \vartheta_0 + \mathcal{O}_{\mathbb{P}}(N^{-\frac{1}{2}-\frac{1}{d}}) \tag{24}$$

as N gets large, i.e. $N^{\frac{1}{2}+\frac{1}{d}}(\hat{\vartheta}_0^N - \vartheta_0)$ is bounded in probability.[5] In particular, the diffusivity can be identified in a finite time horizon T from increasing the spatial resolution. In contrast, the reaction rate ϑ_1 can be recovered as

$$\hat{\vartheta}_1^N = \vartheta_1 + \mathcal{O}_{\mathbb{P}}(N^{-\frac{1}{2}+\frac{1}{d}}) \tag{25}$$

as N gets large, provided that $d \geq 3$. The estimator still converges to ϑ_1, but at a much slower rate. In dimension $d = 2$, the reaction rate can still be consistently estimated, but only with logarithmic convergence rate: $\hat{\vartheta}_1^N = \vartheta_1 + \mathcal{O}_{\mathbb{P}}(\ln(N)^{-1})$. However, in the one-dimensional setting $d = 1$, the estimation error of $\hat{\vartheta}_1^N$ need not even shrink as N grows!

What do the convergence rates (24), (25), which are stated in terms of "spectral resolution" N, tell us about a data frame build of (d-dimensional) pixels? Assume that the side of each pixel has length $h > 0$, such that the size (in pixels) of a frame is of order h^{-d}. Using a discrete Fourier transform, the number of eigenmodes obtained from the data and the number of pixels should be of the same order: $N \sim h^{-d}$. This suggests that according to (24), diffusivity can be estimated at rate $h^{1+d/2}$: For example, in $d = 2$, if the resolution is increased such that h is cut in half, then the uncertainty of $\hat{\vartheta}_0^N$ will decrease by a factor of four. In $d = 3$, the decay is even faster[6]: By making h four times as small we would obtain a decay in uncertainty by a factor of 32. In contrast, by (25), the reaction rate can be estimated only at rate $h^{d/2-1}$: In $d = 3$, if the pixel width is made four times as small, then the uncertainty of $\hat{\vartheta}_1^N$ will decay only by a factor of two! In $d = 2$, the convergence rate is only logarithmic, which will be hard to see in practice.

4.2 Estimation Error Versus Time Horizon

Apart from increasing the resolution within each frame, estimation can be made more precise by observing the trajectory for a longer period of time, i.e. by letting the observation time T be large. It is well-known that drift parameters of stochastic differential equations can be estimated at rate $\mathcal{O}_{\mathbb{P}}(1/\sqrt{T})$, see e.g. [51] for a general discussion. This means that precision is doubled if the observation horizon is extended by a factor of four. There are various works which aim at similar results for stochastic partial differential equations (see e.g. [52–55]). It is important to note that there is no separation between diffusion and reaction terms in the long-time asymptotic regime, i.e. their errors are of the same order if N is fixed and $T \to \infty$. This is a fundamental contrast to the setting of high spatial resolution ($N \to \infty$).

A precise understanding of the convergence rates of the estimators in different asymptotic regimes helps to answer the following question: If I can increase my bud-

[5] In the mathematical literature, this is also called *tightness* and defined as $\lim_{M\to\infty} \sup_N \mathbb{P}(|N^{\frac{1}{2}+\frac{1}{d}}(\hat{\vartheta}_0^N - \vartheta_0)| > M) = 0$.

[6] But of course, cutting h in half in dimension three requires more pixels than in dimension two.

get, should I observe more frames in an experiment or rather increase the resolution within each frame? In the present framework, this turns out to depend heavily on the parameter to be estimated (reaction vs. diffusion).

Strategies to reduce the estimation error of different terms:

	Increase spatial resolution	Extend time horizon
Diffusivity	✓	✓
Reaction terms	(×)	✓

Higher spatial resolution helps to identify reaction terms in $d \geq 2$, but the gain is small (only logarithmic in dimension two).

4.3 Uncertainty Quantification

In certain situations, we have a central limit theorem (CLT). In this discussion, we restrict to the diffusivity estimator $\hat{\vartheta}_0^N$, which should be considered a fundamental quantity due to its stability properties (explained below in Sects. 4.4 and 4.5). Under conditions that we explain below, we have

$$N^{\frac{1}{2}+\frac{1}{d}}\left(\hat{\vartheta}_0^N - \vartheta_0\right) \xrightarrow{d} \mathcal{N}\left(0, \frac{2\vartheta(d+2)}{T\Lambda d}\right) \tag{26}$$

as N gets large, where "$\xrightarrow{d}$" denotes convergence in distribution, and $\Lambda = \lim_{k\to\infty} \lambda_k/k^{2/d}$ is the proportionality constant of the Laplacian eigenvalues on $\mathcal{D}$. This constant can be identified [56, 57] as $\Lambda = 4\pi^2(B_d|\mathcal{D}|)^{-2/d}$, where B_d is the volume of the d-dimensional unit ball, and $|\mathcal{D}|$ is the volume of $\mathcal{D}$ (i.e. the area in $d = 2$). From (26), it follows that for large N, the diffusivity estimator is approximately normally distributed with mean at the true parameter:

$$\hat{\vartheta}_0^N \approx \vartheta_0 + \sqrt{\frac{2\vartheta(d+2)}{T\Lambda d\, N^{1+2/d}}}\mathcal{N}(0,1), \tag{27}$$

and this approximation can be used in a standard way to construct confidence intervals. For example, the interval

$$C_N = \left[\hat{\vartheta}_0^N - 1.96\sqrt{\frac{2\vartheta(d+2)}{T\Lambda d\, N^{1+2/d}}},\ \hat{\vartheta}_0^N + 1.96\sqrt{\frac{2\vartheta(d+2)}{T\Lambda d\, N^{1+2/d}}}\right] \tag{28}$$

is (for large N) an approximate 95% confidence interval.

A simple sufficient condition for the CLT (26) to hold is the absence of a reaction term, i.e. $F_\vartheta(X) = 0$. In this case, $X(t, x)$ is just the solution to a stochastic heat equation. A more interesting condition depends on the stability index of $F_\vartheta(X)$, which we introduce in the next section.

4.4 Misspecification of Reaction Terms

We have seen that reaction terms are intrinsically harder to estimate than diffusivity: Increasing the spatial resolution will give a comparatively small benefit in terms of uncertainty reduction, if at all. This observation, however, has a positive side effect: It goes hand in hand with the fact that the diffusivity estimate is rather *stable* under wrongly specified reaction models.

We can quantify the stability as follows: In the model (4), we associate to $F_\vartheta(X)$ a number $\eta > 0$, called the *stability index*[7] of F. Roughly speaking, η is two minus the order of F as a differential operator. So, a reaction term $f(X)$ has stability index $\eta = 2$, whereas an advection term $\nabla \cdot (Xv)$ has the stability index $\eta = 1$. If $F_\vartheta(X)$ has several components as in (13), its stability index $\eta = \min\{\eta_1, \dots, \eta_p\}$ is just the smallest stability of all its components.

Now assume that the data X is generated from an SPDE that contains the true reaction term F (with stability index $\eta > 0$), whereas $\hat{\vartheta}_0^N$ is constructed from a wrongly specified dynamical model that includes instead the term $\tilde{F}$ (with the same or larger stability index). For example, a local source term may be replaced by another such term (possibly with a different bifurcation structure), or within an advection term, the velocity field may be exchanged. This includes the case where F is just dropped, i.e. formally substituted by zero. Now the question is: What can we say about inference on diffusivity if X comes from the true model, but the estimator $\hat{\vartheta}_0^N$ is built from its misspecified version? The following Theorem 1 now in particular states lower bounds on the stability index, up to which the convergence of $\hat{\vartheta}_0^N$, as well as associated confidence intervals, are not affected.

Theorem 1 ([31])

1. If

$$\eta \geq 1 + \frac{d}{2}, \tag{29}$$

then the convergence of $\hat{\vartheta}_0^N$ as stated in (24) remains unchanged.[8]

---

[7] For a precise mathematical discussion, see [31, Sect. 2.2].

[8] Technically, if (29) holds with equality, the guaranteed convergence rate loses an arbitrarily small polynomial factor, e.g. for $\eta = d = 2$ we have $\hat{\vartheta}_0^N = \vartheta_0 + \mathcal{O}_{\mathbb{P}}(N^{-1+\varepsilon})$ for any $\varepsilon > 0$.

2. *If (29) holds strictly, i.e.* $\eta > 1 + d/2$, *then* $\hat{\vartheta}_0^N$ *even satisfies the CLT (26), leading to identical confidence intervals (28).*
3. *Otherwise, if* $0 < \eta < 1 + d/2$, *then (24) is replaced by*[9]

$$\hat{\vartheta}_0^N = \vartheta_0 + \mathcal{O}_{\mathbb{P}}(N^{-\frac{\eta}{d}}). \tag{30}$$

This has two implications: First, the estimator still converges as N grows. This is a universal property that holds true for any F with positive stability index. Second, if η is large enough as in (29), then not even the convergence rate is affected, i.e. we can still expect the same gain in precision as $N \to \infty$, even if the underlying reaction model is wrong. However, for small η, the misspecified model lowers the rate of convergence. In this case, a larger resolution is needed in order to obtain the same precision as in the correctly specified case.

Interestingly, the stability of $\hat{\vartheta}_0^N$ depends on the dimension d, with better rates in low dimensions. For example, a wrong reaction term with $\eta = 2$ will not affect the convergence speed of $\hat{\vartheta}_0^N$ in dimensions one and two. In $d = 1$, we will even keep the central limit theorem for $\hat{\vartheta}_0^N$ with identical confidence intervals (28), which is explained in Sect. 4.3.

An immediate consequence of the previous discussion is that a wrong advection model affects the diffusivity inference more severely than a wrong reaction model.

In the high resolution limit $N \to \infty$, the diffusivity estimator $\hat{\vartheta}_0^N$ is stable under misspecification in the reaction model.

Note that a similar stability cannot be expected in the large time regime where all estimators will converge at the same rate $\mathcal{O}_{\mathbb{P}}(1/\sqrt{T})$. Thus, increasing the observation horizon will not weaken the impact of model misspecification.

We close this section with a list of commonly used terms and their stability index[10]:

Model for $F(X)$		Stability index
Reaction term	$f(X)$	$\eta = 2$
Advection term	$\nabla \cdot (Xv)$	$\eta = 1$
Inhibitor term as in (8)		$\eta = 4$
Fractional diffusion	$(-\Delta)^{\frac{\alpha}{2}} X$	$\eta = 2 - \alpha$

In align with the discussion in Sect. 4.1, a large stability index should be seen as an indicator that the corresponding term is harder to estimate.

[9] Again, here we neglect an arbitrarily small polynomial factor in the rate.

[10] All examples are discussed in [31, Sect. 2.4] except for the inhibitor term (8), whose stability index is implicit in the proof of [14, Proposition 1].

4.5 Misspecification of Noise Terms

Apart from the reaction term, let us study the robustness of results on $\hat{\vartheta}_0^N$ concerning misspecification of the noise model for ξ. The theory outlined above has been derived under the assumption that ξ is white in time, and possibly colored in space, where the correlation kernel $b_{-\gamma}$ is given by (12). This noise model contains two parameters, namely the noise intensity σ as well as the degree of spatial correlation γ.

The case of wrongly specified parameters. In order to construct $\hat{\vartheta}_0^N$, no knowledge on σ is required. On the other hand, γ does appear in (23). However, if the true exponent γ is replaced by some $\alpha \neq \gamma$ when deriving $\hat{\vartheta}_0^N$, this does not affect any of the results on convergence rates stated above, as long as α is not too small. (For example, one can always choose $\alpha \geq \gamma$.)

We mention that at least in theory, σ and γ can be identified from continuous time observations of X via its quadratic variation. More precisely, for the first eigenmode $X_N = X_1$ of X, given by (15), we have

$$\lim_{k\to\infty} \sum_{i=1}^{k} \int_{\mathcal{D}} \left(X_1\left(\frac{Ti}{k}, x\right) - X_1\left(\frac{T(i-1)}{k}, x\right)\right)^2 \mathrm{d}x = \sigma^2 \lambda_1^{-2\gamma} T, \tag{31}$$

where λ_1 is the first eigenfrequency of the domain $\mathcal{D}$, and the limit refers to convergence in probability. A similar equation holds for the second eigenmode of X (with the second eigenfrequency λ_2 appearing in the limit), and once these two limits are known, the unknown noise parameters σ and γ can be easily recovered.

The case of a wrong parametric model. Next, consider Ornstein–Uhlenbeck noise as in Example 7. Here we induce some slight temporal correlation to the true data generating model. Neglecting this correlation when deriving the estimator turns out to be as severe as a reaction model misspecification ($\eta = 2$). This means, according to (29) and the discussion thereafter, that the convergence rate of $\hat{\vartheta}_0^N$ is not affected in $d = 1$ or $d = 2$, and for $d \geq 3$ it shrinks to the term given by (30). A detailed discussion of this setting can be found in [31, Chap. 3].

4.6 A Counterexample

We have seen that the diffusivity estimator $\hat{\vartheta}_0^N$ is robust against misspecification in the drift, as well as certain deviations from the white noise model. However, this cannot (and does not) apply to all possible kinds of misspecification. A particularly revealing example is the noise model from Example 6, where ξ is modelled as an Ornstein–Uhlenbeck process. We highlight this example, because it is a very natural one: Bearing in mind that the movement of a classical Brownian particle can be described by the Wiener process or an integrated Ornstein–Uhlenbeck process (where the velocity of the particle is modelled as an Ornstein–Uhlenbeck process), the analogous model for ξ would be either space-time white noise, or the model from Example 6. However, if the latter is the true data-generating model for X, this would break the stability of $\hat{\vartheta}_0^N$—in fact, this estimator would not even converge to the true parameter: One can prove that in this setting, even in the simplest case $\mu = 0$ the estimator $\hat{\vartheta}_0^N$ converges to zero as N grows! So what went wrong? The answer is related to the smoothness of X. Note that an integrated Ornstein-Uhlenbeck process has a well-defined derivative, while a trajectory of a Wiener process has not. This observation transfers to our setting: A field X having a well-defined derivative $\partial_t X$ (in the classical sense) is "too regular" to be well approximated by a white noise model.[11]

5 Notes on Implementation

So far, we focused on the situation that the space-time random field X is approximated by some spatially discretized process X_N. In most situations, however, it is reasonable to assume that the data is given as a discrete approximation not only in space, but also in time. In this case, it is straightforward to discretize the integrals appearing in the definition (19) of the least squares estimator $\hat{\vartheta}_N$. There is, however, a caveat for the terms that include the time derivative $\partial_t X_N$, which is related to the fact that mathematically, the field X_N is not differentiable in time. More precisely, consider an integral of the form

$$\int_0^T \int_{\mathcal{D}} F(X)_N \partial_t X_N \mathrm{d}x \mathrm{d}t. \tag{32}$$

Terms of that form arise in (21) or the numerator of (22). These integrals have to be understood in the sense of Itô, which means that they arise as the limit in probability of sums of the form

$$\sum_i \int_{\mathcal{D}} F(X)_N(t_i, x) \frac{X_N(t_{i+1}, x) - X_N(t_i, x)}{t_{i+1} - t_i} \mathrm{d}x \ (t_{i+1} - t_i). \tag{33}$$

Here, it is crucial that the term $F(X)_N(t_i, x)$ is evaluated at the left boundary point t_i of the interval $[t_i, t_{i+1}]$. The reason is that otherwise additional correlation between $F(X)_N$ and the difference quotient is introduced, which will shift the final result.

This is particularly relevant when simulating such integrals (in this case, the mid-point rule or related higher order quadrature methods known from the approximation of a classical Riemann integral do not work without further modification) but also

[11] Note that in this particular example, the problem can be mitigated by replacing X by $Y = \partial_t X$ as the fundamental quantity, i.e. writing down the induced dynamical model for Y and using a least squares method based on that model. See [31, Chap. 3.2] for details.

plays a role when approximating these integrals from discretely given data, always under the standing assumption that the driving noise ξ is (close to) white in time.

In some situations it is possible to avoid the direct approximation of an Itô integral by transforming it with Itô's formula (see e.g. [58]). For example, within the spectral approach (15), the numerator of the "plain diffusivity estimator" (22) can be written as follows:

$$\int_0^T \int_{\mathcal{D}} (\Delta X)_N(t,x)\partial_t X_N(t,x)\mathrm{d}x\mathrm{d}t = \frac{1}{2}\sum_{k=1}^N \lambda_k \left((X_T^{(k)})^2 - (X_0^{(k)})^2 - T\right), \quad (34)$$

where λ_k is the k-th eigenfrequency of the Laplacian Δ on $\mathcal{D}$, and $X_t^{(k)} = \int_{\mathcal{D}} X(t,y)\Phi_k(y)\mathrm{d}y$ is the k-th mode of $X(t,x)$. The right-hand side of (34) allows for explicit calculation. We highlight the Itô correction term $-T$ appearing therein, which is related to correlation appearing in the square of the stochastic process $X_t^{(k)}$.

6 Application to *D. discoideum* Giant Cell Data

We finally show how our method can be applied to measurements of actin concentration within *D. discoideum* giant cell data. The results presented here are taken from [14]. We refer to Sect. 4 therein for a detailed discussion.

We assume that the data follow an activator-inhibitor dynamic as in Example 2. Parametrizing the reaction term as in (14), we have three reaction parameters $\vartheta_1, \vartheta_2, \vartheta_3$ in addition to the unknown diffusivity ϑ_0. The reaction parameters may be treated as known or unknown, leading to a more rigid or flexible statistical model.

We compare four diffusivity estimators, which we apply to the data.

- $\hat{\vartheta}_0^{\text{lin},N}$ is just the estimator (22), where $\vartheta_1, \vartheta_2, \vartheta_3$ are formally set to zero.
- $\hat{\vartheta}_0^{2,N}$ is implicitly defined by (19), where ϑ_0, ϑ_1 are treated as unknown, and the remaining parameters ϑ_2, ϑ_3 are fixed.
- $\hat{\vartheta}_0^{3,N}$ treats $\vartheta_0, \vartheta_1, \vartheta_2$ as unknown and ϑ_3 as fixed.
- $\hat{\vartheta}_0^{4,N}$ treats all parameters $\vartheta_0, \vartheta_1, \vartheta_2, \vartheta_3$ as unknown.

The left panel of Fig. 2 shows the result for a particular measurement of a cell. Interestingly, $\hat{\vartheta}_0^{\text{lin},N}$ and $\hat{\vartheta}_0^{3,N}$ are almost indistinguishable: The detailed reaction model does not seem to contribute to a better understanding of the diffusivity. On the other hand, $\hat{\vartheta}_0^{2,N}$ deviates from the other estimators for low frequencies (in this sense, fixing all reaction parameters may be "too rigid"), but approaches the other estimators for high frequencies, in accordance with the results from Sect. 4.4.

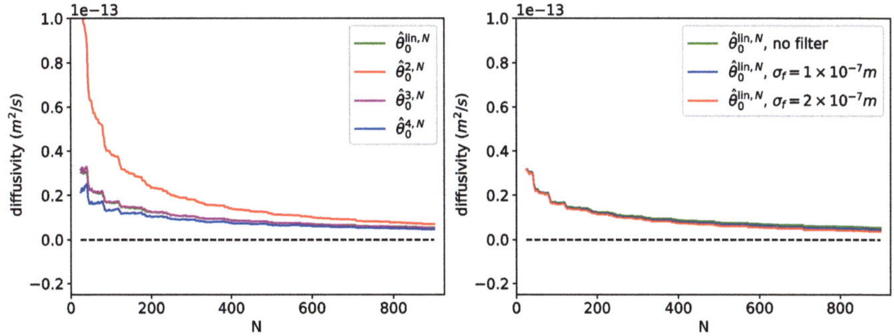

Fig. 2 Plots of $\hat{\vartheta}_0^N$ versus N for different diffusivity estimators $\hat{\vartheta}_0^N$ constructed from experimental data, for $N \geq 25$. *(Left)* Different assumptions are made on what parameters are considered known or unknown. *(Right)* Invariance of diffusivity estimation under preprocessing the data with kernel smoothing in space. Plots reproduced from [14]

One may ask if the assumption that the data is generated by a stochastic reaction-diffusion model is reasonable at all. A first sanity check can be carried out based on the following observation: If the solution to a heat equation is smoothed (in space) with some kernel $K(x)$, the resulting field still solves a heat equation.[12] Consequently, if $X(t, x)$ is generated by a "heat equation up to lower order terms" (which we identify as reaction and noise terms), then $\tilde{X}(t, x) = \int_{\mathcal{D}} K(x - y)X(t, y)\mathrm{d}y$ still comes from such a perturbed heat equation (although the reaction and noise terms will have changed). The right-hand side of Fig. 2 shows the result of applying $\hat{\vartheta}_0^{\mathrm{lin},N}$ to the original cell data as well as to a filtered version of the same data, where a kernel with different values for the bandwidth σ_f is applied as a form of preprocessing. The estimated diffusivity changes only very slightly, as is expected if the stochastic reaction-diffusion model is reasonable.

Next, consider the estimator $\hat{\vartheta}_1^{2,N}$ for ϑ_1, which is constructed from (19) under the model assumptions that ϑ_2, ϑ_3 are fixed and known (in analogy to $\hat{\vartheta}_0^{2,N}$). From (14), ϑ_1 may be identified with the "effective unstable fixed point" of the reaction dynamics. Figure 3 shows the result on simulated and experimental data. On the simulations, increasing the number of Fourier modes N has almost no effect on the value of $\hat{\vartheta}_1^{2,N}$. This is due to the logarithmic convergence rate, as discussed in Sect. 4.1. On real data, the estimated value does indeed lie between zero and one (these two values correspond to the stable fixed points in the FitzHugh–Nagumo system). Further, it is rather stable in N. It may be therefore used as a statistic that reflects an interpretable property of data exhibiting bistability or travelling waves.

---

12 The reason is that the mathematical operations "convolution with a kernel $K(x)$" and "application of the Laplacian operator Δ" commute.

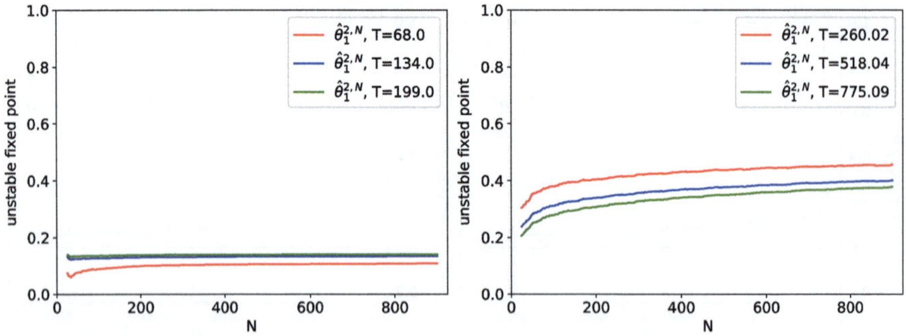

Fig. 3 Graphs of $\hat{\vartheta}_1^{2,N}$ versus N on *(left)* simulated and *(right)* experimental data, where all frames up to time T are used for estimation. All times are given in seconds. We restrict to $N \geq 25$. Plots reproduced from [14]

7 Conclusion

In this work we have discussed statistical inference methods for parameter estimation in stochastic reaction-diffusion systems with a view towards quantitative model evaluation for space-time data $X_N(t, x)$. We discussed the fundamentally different statistical properties of diffusion and reaction estimation, with special emphasis on the issue of drift misspecification. As a proof of concept, we have applied our analysis to the statistical inference on intracellular actin concentration in the social amoeba *D. discoideum*. Both the effective diffusivity and the effective unstable fixed point appear to reflect inherent properties of the data. Both statistics may be used in further research in order to understand and classify spatiotemporal data according to their dynamics.

We hope that the detailed quantitative analysis of the least-squares-based method of parameter estimation, which we have explained in this work, will be helpful in other contexts which we did not consider so far. As a potential example, we mention the pattern formation of zebrafish stripes [59, 60], which can be described by Gierer–Meinhardt type models [61, 62].

So far we did not take measurement noise into account. In the above case of fluorescence microscopy, e.g., this will add independent shot noise and detector noise to $X_N(t, x)$, which requires additional filtering. This will be the content of future work.

Acknowledgements This research has been partially funded by the Deutsche Forschungsgemeinschaft (DFG)—Project-ID 318763901—SFB1294. We like to thank our coauthors from the work [14] Sven Flemming for providing experimental data and Sergio Alonso for sharing numerical code. Plots in Figs. 2 and 3 reproduced from [14] (https://doi.org/10.1007/s00332-021-09714-4, licensed under CC BY 4.0 https://creativecommons.org/licenses/by/4.0/, plots not changed).

References

1. Flemming, S., Font, F., Alonso, S., Beta, C.: How cortical waves drive fission of motile cells. Proc. Natl. Acad. Sci. **117**(12), 6330–6338 (2020)
2. Murray, J.D.: Mathematical Biology II: Spatial Models and Biomedical Applications, 3rd edn. Springer, New York (2003)
3. Edelstein-Keshet, L.: Mathematical Models in Biology: Siam Classics in Applied Mathematics, vol. 46. Society for Industrial and Applied Mathematics, Philadelphia (2005)
4. Meron, E.: Nonlinear Physics of Ecosystems, 1st edn. CRC Press (2015)
5. Kondo, S., Miura, T.: Reaction-diffusion model as a framework for understanding biological pattern formation. Science **329**, 1616–1620 (2010)
6. Beta, C., Kruse, K.: Intracellular oscillations and waves. Annu. Rev. Cond. Matter Phys. **8**(1), 239–264 (2017)
7. Inagaki, N., Katsuno, H.: Actin waves: origin of cell polarization and migration? Trends Cell Biol. **27**, 515–526 (2017)
8. Allard, J., Mogilner, A.: Traveling waves in actin dynamics and cell motility. Curr. Opin. Cell Biol. **25**, 107–115 (2013)
9. Beta, C., Edelstein-Keshet, L., Gov, N., Yochelis, A.: From actin waves to mechanism and back: how theory aids biological understanding. eLife **12**, e87181 (2023)
10. Annesley, S.J., Fisher, P.R.: Dictyostelium discoideum—a model for many reasons. Mol. Cell Biochem. **329**, 73–91 (2009)
11. Vicker, M.G., Xiang, W., Plath, P.J., Wosniok, W.: Pseudopodium extension and amoeboid locomotion in Dictyostelium discoideum: possible autowave behaviour of F-actin. Phys. D Nonlinear Phenom. **101**, 317–332 (1997)
12. Gerisch, G., Bretschneider, T., Müller-Taubenberger, A., Simmeth, E., Ecke, M., Diez, S., Anderson, K.: Mobile actin clusters and traveling waves in cells recovering from actin depolymerization. Biophys. J. **87**, 3493–3503 (2004)
13. Arai, Y., Shibata, T., Matsuoka, S., Sato, M.J., Yanagida, T., Ueda, M.: Self-organization of the phosphatidylinositol lipids signaling system for random cell migration. PNAS **107**, 12399–12404 (2010)
14. Pasemann, G., Flemming, S., Alonso, S., Beta, C., Stannat, W.: Diffusivity estimation for activator-inhibitor models: theory and application to intracellular dynamics of the actin cytoskeleton. J. Nonlinear Sci. **31**(59), 1–34 (2021)
15. Alonso, S., Stange, M., Beta, C.: Modeling random crawling, membrane deformation and intracellular polarity of motile amoeboid cells. PLoS One **13**, e0201977 (2018)
16. Moreno, E., Flemming, S., Font, F., Holschneider, M., Beta, C., Alonso, S.: Modeling cell crawling strategies with a bistable model: from amoeboid to fan-shaped cell motion. Phys. D Nonlinear Phenom. **412**, 132591 (2020)
17. Gerisch, G., Ecke, M., Neujahr, R., Prassler, J., Stengl, A., Hoffmann, M., Schwarz, U.S., Neumann, E.: Membrane and actin reorganization in electropulse-induced cell fusion. J. Cell Sci. **126**, 2069–2078 (2013)
18. Gerhardt, M., Ecke, M., Walz, M., Stengl, A., Beta, C., Gerisch, G.: Actin and PIP3 waves in giant cells reveal the inherent length scale of an excited state. J. Cell Sci. **127**, 4507–4517 (2014)
19. Yochelis, A., Flemming, S., Beta, C.: Versatile patterns in the actin cortex of motile cells: self-organized pulses can coexist with macropinocytic ring-shaped waves. Phys. Rev. Lett. **129**, 088101 (2022)
20. Bressloff, P.: Stochastic Processes in Cell Biology, vol. II. Springer (2023)
21. Lord, G., Powell, C., Shardlow, T.: An Introduction to Computational Stochastic PDEs. Cambridge University Press (2014)
22. Brunton, S.L., Proctor, J.L., Kutz, J.N.: Discovering governing equations from data by sparse identification of nonlinear dynamical systems. Proc. Natl. Acad. Sci. **113**(15), 3932–3937 (2016)

23. Unny, T.E.: Stochastic partial differential equations in groundwater hydrology (Part I: Theory). Stoch. Hydrol. Hydraul. **3**, 135–153 (1989)
24. Kumar, P., Unny, T.E., Ponnambalam, K.: Stochatic partial differential equations in groundwater hydrology (Part 2: Application to Borden aquifer). Stoch. Hydrol. Hydraul. **5**, 239–251 (1991)
25. Altmeyer, R., Bretschneider, T., Janák, J., Reiß, M.: Parameter estimation in an SPDE model for cell repolarization. SIAM/ASA J. Uncertain. Quant. **10**(1), 179–199 (2022)
26. Carrillo, J.A., Craig, K., Yao, Y.: Aggregation-diffusion equations: dynamics, asymptotics, and singular limits. In: Bellomo, N., Degond, P., Tadmor, E. (eds.) Active Particles, vol. 2, pp. 65–108. Springer Nature Switzerland AG (2019)
27. Hübner, M., Khasminskii, R., Rozovskii, B.L.: Two examples of parameter estimation for stochastic partial differential equations. In: Cambanis, S., Ghosh, J.K., Karandikar, R.L., Sen, P.K. (eds.) Stochastic Processes, pp. 149–160. Springer, New York (1993)
28. Huebner, M., Rozovskii, B.L.: On asymptotic properties of maximum likelihood estimators for parabolic stochastic PDE's. Probab. Theory Relat. Fields **103**(2), 143–163 (1995)
29. Altmeyer, R., Reiß, M.: Nonparametric estimation for linear SPDEs from local measurements. Ann. Appl. Probab. **31**(1), 1–38 (2021)
30. Altmeyer, R., Tiepner, A., Wahl, M.: Optimal parameter estimation for linear SPDEs from multiple measurements. arXiv:2211.02496 (preprint) (2022)
31. Pasemann, G.: Parameter estimation for semilinear stochastic partial differential equations. PhD thesis, TU Berlin (2021)
32. Pospíšil, J., Tribe, R.: Parameter estimates and exact variations for stochastic heat equations driven by space-time white noise. Stoch. Anal. Appl. **25**(3), 593–611 (2007)
33. Cialenco, I., Huang, Y.: A note on parameter estimation for discretely sampled SPDEs. Stoch. Dyn. **20**(3), 2050016 (2020)
34. Chong, C.: High-frequency analysis of parabolic stochastic PDEs. Ann. Stat. **48**(2), 1143–1167 (2020)
35. Hildebrandt, F., Trabs, M.: Parameter estimation for SPDEs based on discrete observations in time and space. Electron. J. Stat. **15**(1), 2716–2776 (2021)
36. Hildebrandt, F., Trabs, M.: Nonparametric calibration for stochastic reaction–diffusion equations based on discrete observations. Stoch. Process. Appl. **162**, 171–217 (2023)
37. Kaino, Y., Uchida, M.: Adaptive estimator for a parabolic linear SPDE with a small noise. Jpn. J. Stat. Data Sci. **4**, 513–541 (2021)
38. Liptser, R.S., Shiryayev, A.N.: Statistics of Random Processes I (General Theory). Applications of Mathematics, vol. 5. Springer, New York (1977)
39. Tibshirani, R.: Regression shrinkage and selection via the Lasso. J. R. Stat. Soc. B **58**(1), 267–288 (1996)
40. Hastie, T., Tibshirani, R., Friedman, R.: The Elements of Statistical Learning. Springer, New York (2009)
41. Lototsky, S.V.: Statistical inference for stochastic parabolic equations: a spectral approach. Publ. Mat. **53**(1), 3–45 (2009)
42. Cialenco, I.: Statistical inference for SPDEs: an overview. Stat. Inference Stoch. Process. **21**(2), 309–329 (2018)
43. Huebner, M.: Parameter estimation for stochastic differential equations. ProQuest LLC, Ann Arbor. Thesis (Ph.D.)—University of Southern California (1993)
44. Duncan, T.E., Maslowski, B., Pasik-Duncan, B.: Adaptive control for semilinear stochastic systems. SIAM J. Control Optim. **38**(6), 1683–1706 (2000)
45. Goldys, B., Maslowski, B.: Parameter estimation for controlled semilinear stochastic systems: identifiability and consistency. J. Multivar. Anal. **80**, 322–343 (2002)
46. Cialenco, I., Glatt-Holtz, N.: Parameter estimation for the stochastically perturbed Navier-Stokes equations. Stoch. Process. Appl. **121**(4), 701–724 (2011)
47. Gaudlitz, S., Reiß, M.: Estimation for the reaction term in semi-linear SPDEs under small diffusivity. Bernoulli **29**(4), 3033–3058 (2023)
48. Pasemann, G., Stannat, W.: Drift estimation for stochastic reaction-diffusion systems. Electron. J. Stat. **14**(1), 547–579 (2020)

49. Altmeyer, R., Cialenco, I., Pasemann, G.: Parameter estimation for semilinear SPDEs from local measurements. Bernoulli **29**(3), 2035–2061 (2023)
50. Lototsky, S.: Parameter estimation for stochastic parabolic equations: asymptotic properties of a two-dimensional projection-based estimator. Stat. Inference Stoch. Process. **6**(1), 65–87 (2003)
51. Kutoyants, Y.A.: Statistical Inference for Ergodic Diffusion Processes. Springer Series in Statistics. Springer-Verlag London Ltd. (2004)
52. Koski, T., Loges, W.: Asymptotic statistical inference for a stochastic heat flow problem. Stat. Probab. Lett. **3**, 185–189 (1985)
53. Mohapl, J.: Maximum likelihood estimation in linear infinite dimensional models. Stoch. Models **10**(4), 781–794 (1994)
54. Maslowski, B., Tudor, C.A.: Drift parameter estimation for infinite-dimensional fractional Ornstein-Uhlenbeck process. Bull. Sci. Math. **137**, 880–901 (2013)
55. Kříž, P., Maslowski, B.: Central limit theorems and minimum-contrast estimators for linear stochastic evolution equations. Stochastics **91**(8), 1109–1140 (2019)
56. Weyl, H.: Über die asymptotische Verteilung der Eigenwerte. In: Nachrichten von der Gesellschaft der Wissenschaften zu Göttingen, Mathematisch-Physikalische Klasse, pp. 110–117 (1911)
57. Shubin, M.A.: Pseudodifferential Operators and Spectral Theory, 2nd edn. Springer, Berlin, Heidelberg (2001)
58. Gall, J.-F.L.: Brownian Motion, Martingales, and Stochastic Calculus. Springer (2016)
59. Yamaguchi, M., Yoshimoto, E., Kondo, S.: Pattern regulation in the stripe of zebrafish suggests an underlying dynamic and autonomous mechanism. Proc. Natl. Acad. Sci. **104**, 4790–4793 (2007)
60. Müller, P., Rogers, K.W., Jordan, B.M., Lee, J.S., Robson, D., Ramanathan, S., Schier, A.F.: Differential diffusivity of nodal and lefty underlies a reaction-diffusion patterning system. Science **336**, 721–724 (2012)
61. Gierer, A., Meinhardt, H.: A theory of biological pattern formation. Kybernetik **12**, 30–39 (1972)
62. Koch, A.J., Meinhardt, H.: Biological pattern formation: from basic mechanisms to complex structures. Rev. Mod. Phys. **66**, 1481–1507 (1994)

Networks and Interactions

Networks: The Visual Language of Complexity

Blai Vidiella, Salva Duran-Nebreda, and Sergi Valverde

Abstract Understanding the origins of complexity is a fundamental challenge with implications for biological and technological systems. Network theory emerges as a powerful tool to model complex systems. Networks are an intuitive framework to represent inter-dependencies among many system components, facilitating the study of both local and global properties. However, it is unclear whether we can define a universal theoretical framework for evolving networks. While basic growth mechanisms, like preferential attachment, recapitulate common properties such as the power-law degree distribution, they fall short in capturing other system-specific properties. Tinkering, on the other hand, has shown to be very successful in generating modular or nested structures 'for-free', highlighting the role of internal, non-adaptive mechanisms in the evolution of complexity. Different network extensions, like hypergraphs, have been recently developed to integrate exogenous factors in evolutionary models, as pairwise interactions are insufficient to capture environmentally-mediated species associations. As we confront global societal and climatic challenges, the study of network and hypergraphs provides valuable insights, emphasizing the importance of scientific exploration in understanding and managing complexity.

Keywords Networks · Evolution · Hypergraphs · Complex systems · Tinkering

B. Vidiella
Theoretical and Experimental Ecology Station, CNRS, Moulis, France
e-mail: blaivr.lab@gmail.com

S. Duran-Nebreda · S. Valverde (✉)
Institute of Evolutionary Biology, CSIC-UPF, Barcelona, Spain
e-mail: sergi.valverde@ibe.upf-csic.es

S. Duran-Nebreda
e-mail: salva.duran@ibe.upf-csic.es

S. Valverde
European Centre for Living Technology (ECLT), Ca' Bottacin, Venice, Italy

© The Author(s), under exclusive license to Springer Nature Switzerland AG 2025
M. Stich and J. Carballido-Landeira (eds.), *Nonlinear Dynamics for Biological Systems*,
SEMA SIMAI Springer Series 40,
https://doi.org/10.1007/978-3-031-99044-1_6

1 Introduction

Understanding complexity is a main scientific challenge [16]. This quest has been a complex process in and of itself, motivated by both theoretical and practical reasons. Consider two examples of using networks to comprehend complexity, one from biology and one from technology. Bertrand Russell proposed in the 1930s that notations play an important role in distilling abstract ideas into clearly intelligible forms, hence aiding communication [124]. Almost 90 years later, the Synthetic Biology Open Language (SBOL) [63] introduced a formal notation in modern biology, aspiring to be an industrial standard for those engineering or simply conveying biological designs (particularly genetic constructs and pathways) (see Fig. 1a).

Prior to SBOL, computer scientists were seeking for a language that would allow them to control the complexity of software. Alan Kay, a former math and biology student and Xerox PARC computer scientist, recognized that the level of complexity in software was only comparable to that of biology. Kay played a significant role in the development of Smalltalk [55], a programming language that conceptualized

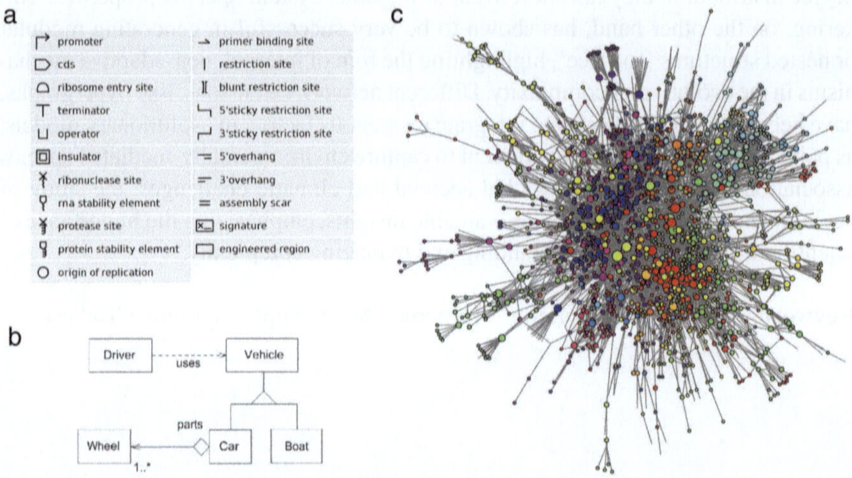

Fig. 1 Visual notations in biology and software. **a** The Synthetic Biology Open Language (SBOL) defines standardised visual notations for representing genetic constructions and biological designs such as DNA sequences, regulatory elements, and functional parts, as well as their connections and interactions within a biological system. Visual symbols represent promoters, coding sequences, terminators, and other genetic components, making it easier to communicate biological ideas across communities. **b** In software engineering, class diagrams are a form of structural diagram implemented in the Unified Modeling Language (UML). They serve the purpose of conceptualizing and representing the distinct attributes of software components, as well as their interrelationships, within a given system or software application. The illustration depicts a section of the class diagram used to create a racing game. **c** The network representation of the class diagram in the video game Prorally 2002 (Ubisoft, 2002). Colour denotes the subsystem the software component belongs to (i.e., physics, rendering, audio, etc.), and node size represents the in-degree

software as a network of code units, drawing inspiration from biological analogies and the interactions between cells: "I thought of objects being like biological cells and/or individual computers on a network, only able to communicate with messages."

Smalltalk is a successful programming language, but neither it nor its contemporary descendants have been the ultimate solution to software development [106]. One of us participated in a video game project in 2002, where we used software engineering techniques, including graphical notation (see Fig. 1b). The large-scale structure of our video game proved to be extremely complex and difficult to understand, despite the talent of the software development team and ongoing attempts to optimize, improve, and modularize the software architecture [107, 113] (see Fig. 1c). Unwanted complexity is certainly not surprising to many software developers, but neither is there a convincing explanation for it (or a strategy to prevent it). For these reasons, we decided to launch an ongoing research project to investigate the origins of software complexity from a scientific (not engineering) perspective [110]. This example also suggests that SBOL's comparable initiatives may have a limited impact in managing biological complexity (especially when dealing with large non-orthogonal biological designs [22, 23]). But why are notation schemes insufficient? What is missing?

An important requirement is the ability to capture the structure-dynamics relationship as well as the evolutionary component inherent in complex systems. The crucial role of dynamics in complex systems is exemplified by social insect colonies, whose properties cannot be reduced to the agents building them [123]. Nest architecture emerges from the coordinated actions of many insects, which are, in turn, channeled through the emerging 3-D network of galleries [75, 108]. Feed-backs are a universal feature of the evolution of complexity, which can be a driver of adaptations and innovations emerging at multiple scales [30]. Examples include not only social insects [88, 123], but multicellularity [19, 31, 78, 89], human language [26] and many others [104]. Evolutionary transitions frequently occur at critical points [87, 99] that often involve (or can be understood by) some kind of structural property. Network theory is a tool for not only graphically displaying the structure of a complex system, but also for understanding the logic and evolutionary principles that underpin these systems.

Networks provide a means to investigate the relationship between system structure and dynamics, offering an intuitive framework to represent intricate interdependencies among system components. Their versatility lies in their ability to capture both local and global properties, enabling the analysis of small-scale interactions and large-scale emergent phenomena. Whether it's understanding the spread of diseases, analyzing information flow in social media, or optimizing transportation networks, graphs offer a robust mathematical language for studying complex systems and their dynamics. The effectiveness of networks as a mathematical representation of complexity has led researchers to embrace them in various scientific disciplines. This multidisciplinary approach facilitates the integration of theoretical and empirical approaches, allowing us to better comprehend and predict abrupt human-induced changes in ecosystems. Many examples demonstrate how networks can deepen our understanding of the impact of human actions on the biosphere. By leveraging the power of networks, we can develop strategies to mitigate the effects of human actions, a crucial challenge in the era of climate change.

2 Measures

A first step in analyzing the structure of complex systems is often to make a picture of it. The human eye is extremely powerful at picking out visual patterns and allow us to put visualization to work on our studies. A network $G = (V, E)$ is a graphical representation that consist of just two fundamental elements: nodes (or vertices) $v_i \in V$ and pairs of nodes $(v_i, v_j) \in E$ so-called links or edges. Despite this apparent simplicity, networks can express complex patterns by connecting nodes and links in numerous ways. This approach may be beneficial even in systems that do not inherently suggest a network [60] (see Fig. 2).

Network visualization is, however, only really useful for (a) small networks (up to a few hundred or thousands of vertices), and for (b) sparse networks, that is, when the density of connections is low as not to obscure the main structural features. If there are too many vertices or links, then renderings will be complicated and difficult to understand (such network diagrams are popularly known as 'hairballs'). Many of the networks that interest scientists today have hundreds of thousands or even millions of nodes, which means that visualization is not of much help in their analysis and

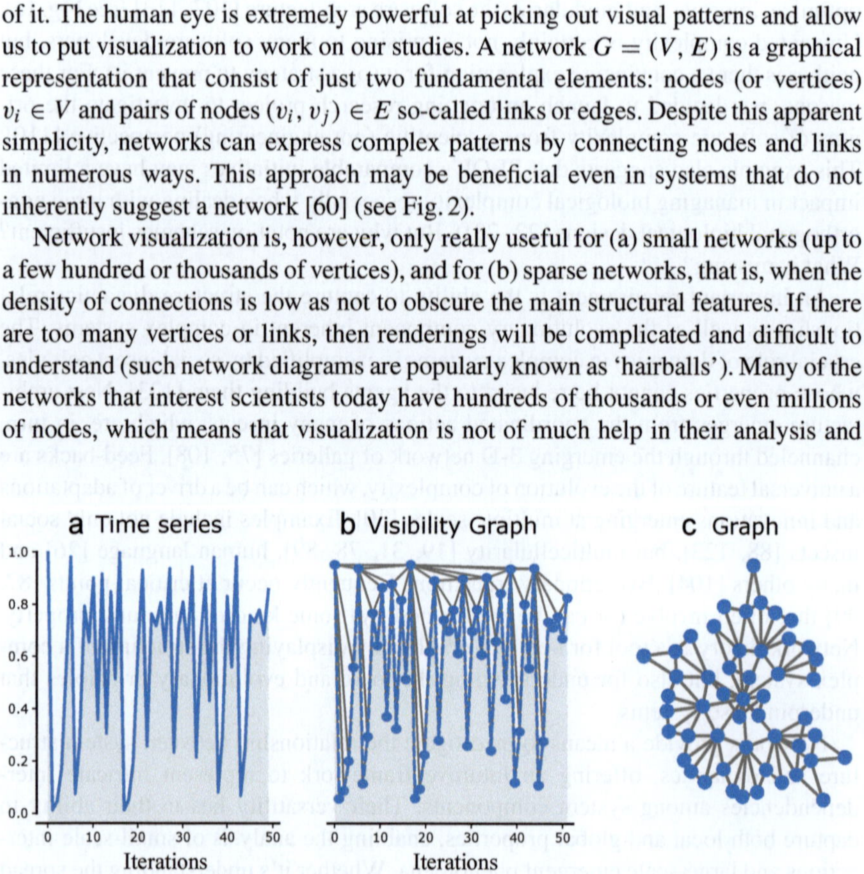

Fig. 2 Network representation of time series. **a** A Horizontal Visibility Graph (HVG) is a graph representation constructed from time series or one-dimensional data. The HVG is formed by converting time series data into a graph using an algorithm based on the idea of horizontal visibility. **b** Each time series point is treated as a node in the HVG. When a data point at one time step "sees" another data point at a different time step and no other point between them has a higher value, a link is established. **c** The HVG enables the use of graph theory to understand the underlying structure or features of time series data in a variety of domains such as signal processing, economics, and natural sciences

we need to employ other techniques to determine their structural properties. Here, network theory has developed a large toolkit of measures and metrics that can be useful to understand what data are telling us, even in cases where visualization is not possible.

2.1 Degree

A convenient notation for mathematical purposes is the adjacency matrix. The adjacency matrix A of a simple network $G = (V, E)$ is the matrix with elements $A_{ij} = 1$ iff $(i, j) \in E$ and $A_{ij} = 0$ otherwise. The structure of the adjacency matrix is informative. For an adjacency matrix with no self-edges, the diagonal elements are all zero, while a symmetric matrix shows that the network edges are all bi-directional (so-called 'undirected' edges). In some cases, we would like to represent edges as having a strength, or a weight, or a value attached to them. For example, Internet edges might have weights representing the amount of data flowing along them. In a social network, edges might have weights representing the frequency of interactions. In neural networks, learning processes reinforce specific connections between neurons, and weights might represent the likelihood of signal propagation from one neuron to the next. These weighted networks can be represented by giving the elements of the adjacency matrix values equal to the weights of the corresponding connections, i.e., the volume of traffic carried by a communication link, the number of social interactions, or the strength of a neuronal connection.

The most basic structural property of any node i is its degree $k(i)$ or the number of links connected to it. The degree can be written in terms of the adjacency matrix as follows:

$$k_i = \sum_{j=1}^{N} A_{ij} \tag{1}$$

where $N = \|V\|$ is the number of nodes. Every edge of an undirected network has two endpoints and if there are m edges in total there are $2m$ ends of edges:

$$2m = \sum_{i=1}^{N} k(i) \tag{2}$$

or

$$m = \frac{1}{2}\sum_{i=1}^{N} k(i) = \frac{1}{2}\sum_{i=1}^{N}\sum_{j=1}^{N} A_{ij} \tag{3}$$

The above result is often used in the study of networks. Notice that the maximum possible number of edges in an undirected network is

$$\binom{N}{2} = \frac{1}{2}N(N-1) \tag{4}$$

By dividing the number of edges m by this maximum possible, we obtain the connectance (ρ) or density of a network, i.e., the fraction of edges that are actually present [76]:

$$\rho = \frac{m}{\binom{N}{2}} = \frac{2m}{N(N-1)} = \frac{\langle k \rangle}{N-1} \tag{5}$$

where $\langle k \rangle$ is the average degree of a vertex in an undirected network. A dense network is one in which the density tends to a constant as N approaches infinity. In this situation, as the network grows in size, the proportion of non-zero entries in the adjacency matrix remains constant. The density of a sparse network, on the other hand, approaches zero as the network size increases, and so the average degree tends to remain constant. Almost all real-world systems are thought to be sparse networks, which has profound implications for our understanding of complex systems [90].

2.2 *Percolation*

A system is a set of nodes that are linked together through a chain of intermediate connectors. Nodes that have no connection to the rest of the connected group of species, genes, or people are almost certainly to be removed from the global structure. This raises the question of how the system components were first joined together.

A simple model provides a broad view of this problem (see Fig. 3). Consider a square grid of $L \times L$ water pipes. This spatial network has three types of pipes: (1) pipes that are only open, (2) pipes that are closed, and (3) pipes that are open and may be linked to the left-hand side via a global route. We are interested in the circumstances that allow global routes to appear. Assume that a random pipe on this lattice is opened with probability p, with the other pipes remaining independent. Open pipes look separated from one another with low p values. Although the density of open pipes rises linearly with p (on average, we will have pL^2 open pipes), the global connectedness does not. The fraction of connected pipes is a non-linear function of the probability p. For low p, this fraction stays constant until the critical value of $p = 1/2$ (the so-called 'percolation threshold') is achieved, at which time the water percolates the grid and a global route connects its opposing sides [56].

Percolation transitions like these occur in all types of networks. The simplest model of a random network, the Erdős-Rényi graph [39], provides a null hypothesis for percolation phenomena in disordered systems. The number of nodes N in the random graph is fixed, and each edge is present with probability p and missing with probability $1 - p$. Consider the two extreme scenarios of this model. When $p = 0$, each vertex is an island: the network comprises N isolated components, each of which has precisely one vertex. When $p = 1$, the random network shows all

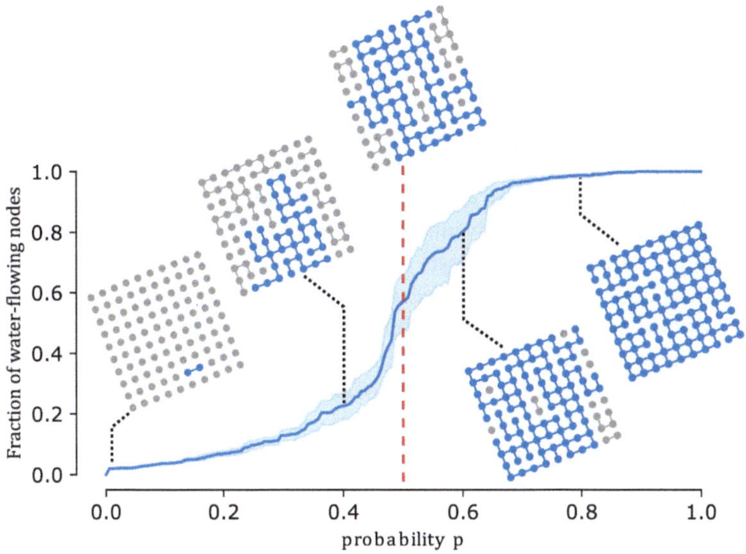

Fig. 3 Using a square regular lattice for water transportation. In a pipe grid, the probability p (horizontal axis) of opening a pipe (forming an edge) and the total fraction of nodes transporting water (vertical axis) are not linearly related. Nodes on opposing sides of the lattice are part of global transportation routes beyond a critical probability of $p > p_c = 1/2$, even when pipes are randomly opened. Blue and grey circles represent water-flowing and disconnected nodes, respectively

possible edges, resulting in a single linked component. A 'clique' is the formal term for this configuration in which a network of n nodes is linked to every other node.

What happens when probability of connection continuously increases from $p = 0$ to $p = 1$ in a random network? We may hypothesize that the size of the largest connected component $S(p)$ grows linearly with p. But something much more interesting happens. The fraction S of nodes in the largest connected component changes abruptly as the system progresses from disconnected to fully connected. At some critical value of the average degree $z = p(N - 1)$, the random network exhibits a phase transition [87]. A large connected component emerges at the critical point $z = z_c = 1$. When $z > z_c$, most nodes are linked in a single graph; however, when $z < z_c$, the system breaks down into many little subgraphs. That is, the random network predicts the emergence of a percolation threshold separating a disconnected phase from another configuration in which most (if not all) components are connected by at least one global route. This prediction applies to all types of networks, as the local structure at the percolation threshold is expected to be a tree with no loops [21]. However, in real systems, there is also a tendency to short loops [77], which are also very sparse structures.

2.3 Clustering

Random graphs explain the basic features of real-world networks, like the emergence of a large connected component associated with the percolation transition. The random network must be extended to account for other empirical features, particularly when examining local network structures. For example, in social networks [74], the notion of transitivity significantly influences the associations between nodes [49].The ideal concept of transitivity suggests that connections follow a 'friend of a friend' pattern, often observed in well-connected, fully-linked networks (or cliques). Yet, most real-world networks, being inherently sparse, operate under partial transitivity. For instance, the fact that 'A' is acquainted with 'B' and 'B' with 'C' does not guarantee that 'A' directly knows 'C' in social networks.

The clustering coefficient may be used to measure the degree of partial transitivity in real systems. The clustering coefficient c_i of a vertex i is the ratio between the number of edges e_i among its nearest neighbors and its maximum possible value [70]:

$$c_i = \frac{e_i}{\binom{k_i}{2}} = \frac{2e_i}{k_i(k_i - 1)} \tag{6}$$

This coefficient ranges from 0 to 1. $c_i = 1$ denotes complete transitivity, i.e., a network composed entirely of cliques. If $c_i = 0$, there are no closed triads. This can happen in a number of network types, including trees (which do not have any loops) and square lattices (which have closed loops with an even number of vertices but no closed triads). There is no transitivity or network clustering in random networks because the chance that any two vertices are neighbors is the same, no matter what the nodes are. As a result, in a random network, the average clustering coefficient is $\langle c \rangle = p = \langle k \rangle/(N - 1)$, which goes to zero in the limit of high N. Empirical clustering values often range from 0.01 to 0.5, indicating considerable differences between random and real networks [118].

Many real-world networks display an approximate degree dependence, with higher degree vertices having a lower clustering coefficient on average [64]. Clustering may also be used to detect so-called "structural holes" [20] in the network, or missing linkages between unrelated nodes linked indirectly via a third vertex. Structured holes are negative for transmission efficiency because they restrict the number of alternate pathways through which information may travel. Structured holes, on the other hand, might be advantageous for a central node whose friends lack connections, granting control over information flow between those friends.

2.4 Motifs

Clustering coefficients are instrumental when looking at the local structure of social networks. When applied to directed networks, where imbalanced information, energy, or matter flow predominates, this coefficient is less effective. An alternative lens to explore network architecture involves network motifs [66]. These motifs denote compact subgraphs comprising M nodes, interconnected by a subset of links within a larger network with $N > M$ nodes. For each value of M, a finite set of variants exists (see Fig. 4a). Counting the occurrences of subgraphs within a network generates a frequency spectrum that provides key insights into their role as basic building blocks. Some of these subgraphs occur at considerably greater frequencies than expected in random graphs. To assess the significance of these occurrences, the Z-score provides a statistical metric that quantifies the prevalence of a particular subgraph Ω_i across the network [66]:

$$z(\Omega_i) = \frac{N_{real}(\Omega_i) - \langle N_{rand}(\Omega_i) \rangle}{\sigma(N_{rand}(\sigma_i))} \tag{7}$$

where N_{real} is the number of appearances of the subgraph Ω_i in the network, $\langle N_{rand} \rangle$ is the average number of appearances of the subgraph Ω_i in a large number of randomised networks with the same number of nodes and degree distribution as the original network, and σ denotes the standard deviation. A subgraph's Z-score may be positive or negative; $z(\Omega_i) > 0$ indicates that it is greatly overrepresented (it is a motif) in the original network relative to randomized, and $z(\Omega_i) < 0$ indicates that it is severely underrepresented (anti-motif). There are some subgraphs that are overrepresented in the structural and functional networks of the human brain [68, 95].

Although motif analysis is a valuable tool, e.g., when assessing the vulnerability of complex networks [28], its biological relevance in general remains unclear [58, 91]. However, synthetic biologists have made extensive use of the motif framework to understand and exploit design spaces in their biological designs. For instance, an exhaustive analysis of feedforward motifs in regulation has revealed that the most common 'band-pass filter' design in evolution is also the most robust one, both in terms of dynamical properties as well as resilience to mutations in their components [82]. Similarly, understanding biological motifs underpins our capacity to engineer self-organized criticality [117], associative learning [62] and more generally, both temporal [36] and spatial patterns [33] (see Fig. 4b, c).

Still, it would be unreasonable to directly infer functionality from the presence or absence of specific motifs, given that a specific subgraph might fulfill a variety of roles in different systems. For example, flexible components that modify their mode of operation to replace damaged or absent components might achieve resilient functionality [101]. According to synthetic models, brain networks are formed to maximize the number and diversity of motifs, as local structural variability enables an extensive spectrum of functional states [95]. Computational research suggests that the degree of synchronization exhibited by a motif is influenced by its structure.

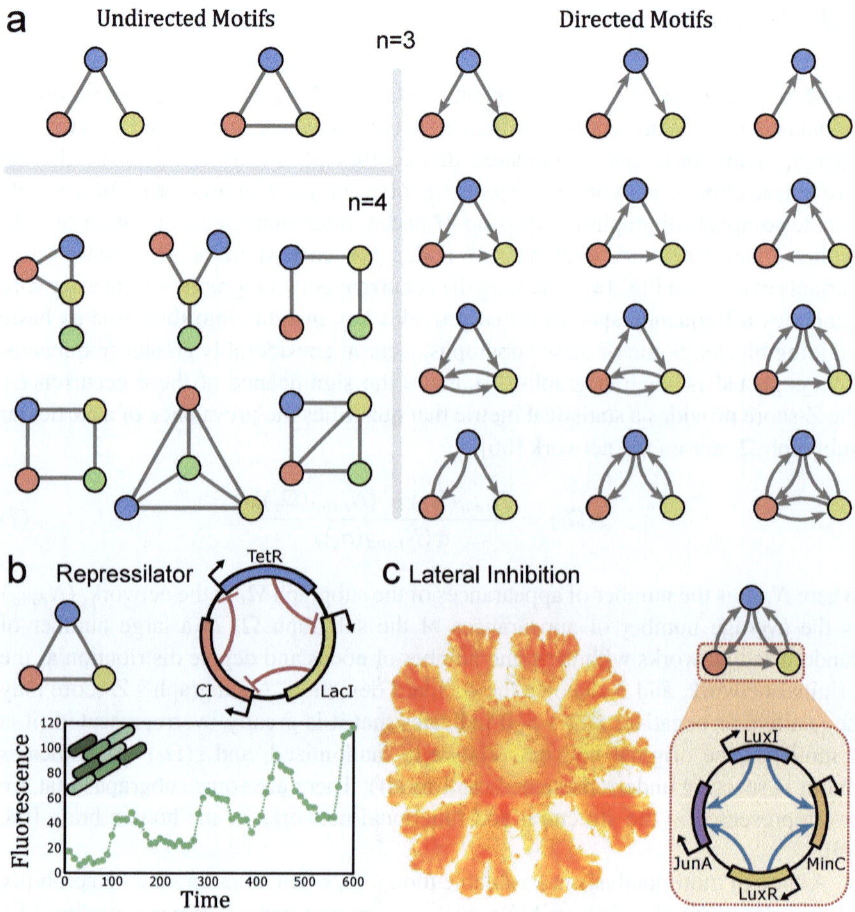

Fig. 4 Motif analysis in biology. **a** List of 3 and 4 node undirected network motifs. Directed motifs (right) display a wider range of patterns than undirected motifs (left), resulting in more complex networks of connections. Synthetic biologists have taken inspiration from these smaller subnetworks to design complex biological circuits. These motifs have proven useful in the implementation of temporal patterns, such as **b** oscillators, e.g., the repressilator circuit [36], which features a cascading triad of inhibiting transcriptional regulators (typically TetR, CI and LacI), yielding an oscillation every 150 min. The repressilator can be exogenously controlled and owes its robust behaviour despite its simplicity to the underlying transcriptional motif. Robust spatial pattern formation induced by lateral inhibition **c** also hinges on specific design motifs [33], in the example shown a communication-coordinated feedback loop (LuxR-LuxI) promotes cellular adhesion (JunA) while also arresting bacterial growth (MinC), creating regular patterns of cellular density with a characteristic wavelength of 0.2 cm in synthetic *E. coli* colonies

Motifs with greater density of links are more likely to synchronize compared with those with fewer connections.

3 Evolution

While random graphs offer insights into network structures, they possess a fundamental limitation in their static nature. Random theory depict networks where all nodes are simultaneously created and remain unaltered, overlooking the dynamic expansion processes prevalent in real-world systems. Take cities, for example. They exhibit a dynamic nature akin to living organisms [119, 120]. The complex development of urban networks has been studied from the perspective of fractals [15] and chaos theory [24], since both include self-similar structures as well as sensitivity to initial conditions. This complex city expansion often reflects a combination of distinct developmental patterns: an urban network core surrounded by a grid-like expansion that conforms to the principles of urban planning, while the periphery experiences a more organic and unpredictable growth [114]. A static depiction of networks overlooks the complex interplay among historical events, regulatory influences, and selection factors. Embracing dynamical models is crucial to understanding the evolving dynamics of cities [10] and other complex networks.

An open question is to what extent we can define a universal theoretical framework for evolving networks. While basic growth mechanisms, like preferential attachment, contribute to explaining common properties such as the power-law degree distribution, they fall short in capturing the distinctive nature of individual networks. Understanding the evolution of real-world networks necessitates a combination of structural traits beyond local features.

3.1 Growth

Many interesting examples of growing networks are related to evolutionary processes, whether in natural or artificial systems. For example, although technical breakthroughs are frequently the result of purposeful efforts, George Basalla observed that a broad range of inventions cannot be explained solely by intentional processes, implying deep analogies between biology and technology that have been the focus of extensive research [11, 72, 97]. Furthermore, networks may be used to explore the parallels (as well as the differences) across various evolutionary systems. Historical events both have a significant impact on the course of technology and biology, often resulting in unexpected detours on the path to innovation [5]. Biology and technology also depend on differential replication, in which accidental, seemingly random changes are inherited by descendants. Similar to selection forces in evolution, artifacts are also susceptible to external success measures [65], such as reliability, performance, novelty [32, 69], and even aesthetic preferences.

Networks are a promising approach to study the evolution of technology, which lies at the intersection between statistical physics, computer science, and evolutionary biology [93]. The network of patent citations, for example, is a scale-free network, a property also observed in other physical, natural, and man-made systems in which the degree distribution $P(k)$ is a power-law rather than a normal distribution:

$$P(k) = Uk^{-\gamma} \tag{8}$$

where U is a constant, and $2 \leq \gamma \leq 3$ is the exponent of the power-law [8]. Power law distributions are different from normal distributions in that they do not have a peak at the average degree $\langle k \rangle$, and they are more likely to contain extreme degree values.

Assuming that the amount of citations is a reliable indicator of a technology's importance, the power-law distribution of citations may be interpreted as an indication of inequality between innovations. But what factors contribute to the popularity of a node? De Solla Price suggested the 'cumulative benefit' mechanism for citation networks in 1965, stating that 'there is a likelihood that the more an article is cited, the more probable it is to be mentioned later' [94]. This process is similar to the 'rich get richer phenomenon' proposed by Yule [125] and Simon [85], as well as the newer concept of 'preferential attachment' [9].

We can incorporate the mechanism of 'cumulative advantage' into a model of growing networks, where the rate of incoming links is proportional to the degree of any node:

$$\Pi(k) \sim k^{\beta} \tag{9}$$

where the attachment exponent β regulates the strength of link reinforcement. It is possible to demonstrate that the above equation produces a degree distribution that follows a power-law [9, 125]. Although this model is often simplistic for reproducing real-world degree distributions, it can be easily expanded. For example, this equation does not take into account existing temporal relationships between inventions. To understand the evolution of technology, we must consider that individual traits are not the unique criteria of success; the environment must also be adequate, since their role may change over time in evolution (e.g., an exaptation [48]). Taking into account this, we can define a more realistic model for the growth of patent citations that combines preferential attachment with temporal dependence [115]:

$$\Pi(k, \tau) \sim k^{\beta} \tau^{\alpha-1} \exp(\tau/\tau_0)^{\alpha} \tag{10}$$

where τ is the node age, the exponent $\alpha > 0$ weights how quickly the ageing influences the chance of attachment, and τ_0 is a scale parameter that determines the rightward extension of the ageing curve. We can verify these assumptions in a database of over 3 million patents [115], which suggests a complex interplay between structure and environment in the evolution of technology. However, a power-law degree distribution is a general feature that many systems have, hence it cannot identify what distinguishes a network from others.

3.2 Modularity

A modular system is formed by quasi-independent parts that appear integrated within themselves, but also exhibit a certain degree of interdependency among them. Modularity is considered a prerequisite for the adaptation of complex organisms and their evolvability [83]. This is particularly evident in cellular networks, where modularity may be identified at the topological level [79, 92]. In many situations, modular architectures seem to be linked to functional traits: for example, a group of closely similar proteins may all be involved in cell division or communication [53]. While empirical research, e.g. [38], suggests a potential relationship between modularity and functional traits, no universal laws or consistent patterns have emerged in relation to modular structure. A main obstacle is the need to get reliable functional data, which involves the detailed and costly inspection of a large number of nodes.

An active area of research combines statistical physics with computational approaches to develop community detection algorithms that solve this problem [70]. The approach takes into account a decomposition of the graph $G = (V, E)$ into a collection of s subgraphs $C = \{C_i \subset V : 1 \leq i \leq s\}$ that defines a partition. Because there are so many possible partitions, it is critical to have a reliable index for assessing the degree of connectivity between alike nodes in the network. The optimal partition is determined by maximizing the modularity index $Q = Q_{\max} = \max(Q)$ [71]. The modularity score is:

$$Q = \sum_{i=1}^{s} \frac{e_i}{m} - \left(\frac{d_i}{2m}\right)^2 \tag{11}$$

where e_i is the number of edges in module i, d_i is the total degree of nodes in module i and m is the number of edges in the full network. This method is widely used, but it comes with several limitations (see Fig. 5). The so-called 'resolution limit' [42] is a significant constraint that forces the algorithm to integrate small modules whenever the merging gives a positive gain. Let's define the gain ΔQ as follows:

$$\Delta Q_{ij} = \frac{e_{ij}}{m} - 2\left(\frac{d_i}{2m}\right)\left(\frac{d_j}{2m}\right) \tag{12}$$

where e_{ij} is the number of edges between modules i and j. A positive gain $\Delta_Q > 0$ occurs when the number of edges between two subsets of nodes is greater than the expected number in a random graph [46]:

$$e_{ij} > \frac{d_i d_j}{2m} \tag{13}$$

In sparse or large networks, this value can fall below one ($e_{ij} < 1$), prompting the merging of weakly linked modules, contrary to our intuition that they should remain separate [46]. The resolution limit results from the modularity equation's null model, which assumes equal probabilities of node connections. This assumption may not hold valid in large or geographically embedded networks, as each node may only have a finite "horizon" within which it may communicate with other nodes [111]. For example, each connection in an intercommunication network has a cost that lowers the likelihood of connecting with another one separated by a large distance [25].

Modularity maximization presents an inherent challenge: maximizing the first term of the modularity equation often requires including numerous edges within modules, while reducing the second term necessitates dividing the graph into smaller, low-degree modules. Finding the optimal balance is a hard computational problem. And as a result, all practical modularity algorithms are based on heuristics, which yield suboptimal partitions [18]. Furthermore, it is unclear how to tune the parameters of the maximization algorithm to obtain the optimal solution. One method involves sweeping the parameter space by varying the temperature in a simulated annealing search [84]. The network gets clustered into an increasing number of groups and modularity values when the temperature is lowered. The shift from low to high modularity is not smooth, but consists of multiple phases separated by abrupt discontinuities (see Fig. 5a). The presence of multiple potential solutions, each with slightly different modularity scores for a given network (Fig. 5b), reveals a landscape of local maxima [46] (see Fig. 5c). To address this problem of degeneration, consensus-based solutions that identify resilient communities are necessary.

Evolution cannot invariably produce maximally modular networks when functional and cost constraints are present, which is a more significant limitation than the constraints associated with a mathematical definition of modularity. The so-called "breakdown of modularity" involves a transition from modular to well-connected networks [43, 105], and it could be related with neurological disorders including Alzheimer's disease or schizophrenia [2, 27]. An extensive search of the space of 3-input logical functions reveals that not all minimal feed-forward Boolean networks (FFBNs) are modular [105] (see Fig. 5d). Looking at this phenotypic network, we can observe that evolution of circuits such as the multiplexer and the majority function, which are important in both electrical circuits and synthetic biology, is linked to a modularity reduction from an ancestral Boolean function. In these circumstances, the only way to retain modularity is to raise the cost of the circuit, for example, by increasing structural redundancy.

3.3 Tinkering

Complex systems emerge from the nonlinear interactions among numerous components, leading to unpredictable behaviors. Modular structures are a frequent component of many complex systems, both in nature and in engineering. In technology, modularity provides significant benefits as it establishes distinct task divisions, aim-

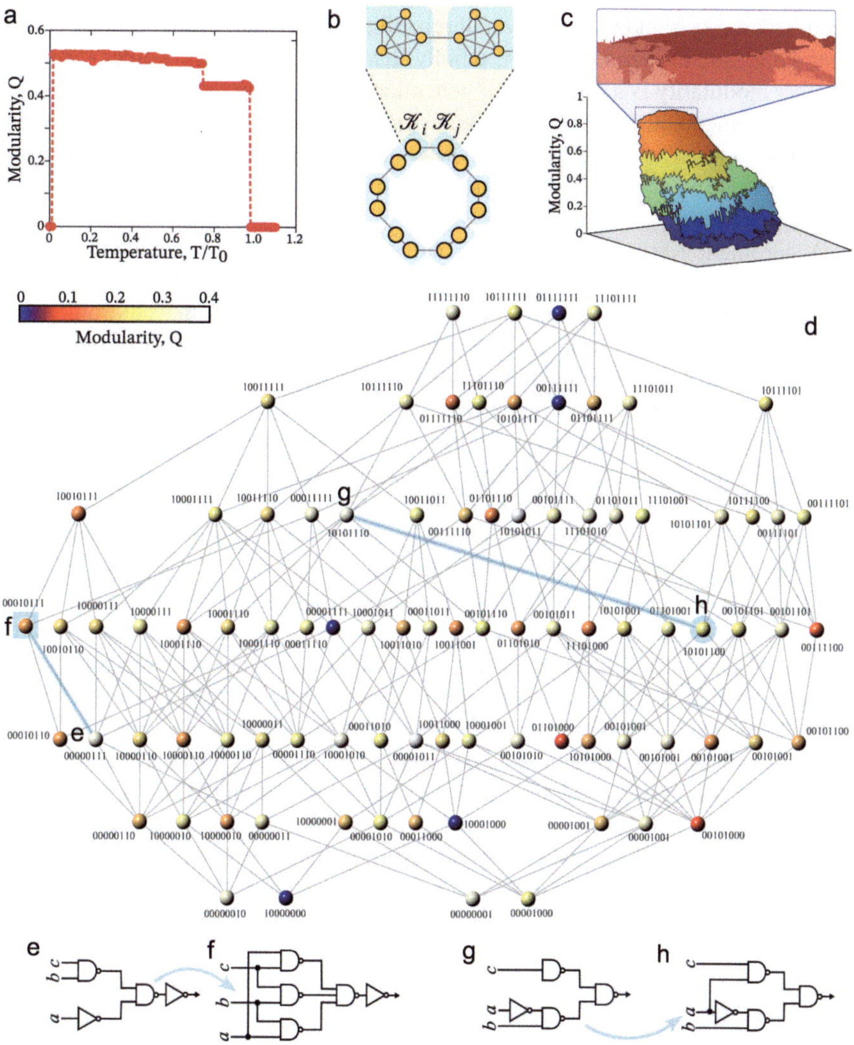

Fig. 5 Limits to modularity. **a** A simulated annealing search across a broad temperature range finds abrupt modularity transitions separating homogeneous phases. The presence of stable phases suggests how a partial (heuristic) search is likely to provide sub-optimal modularity values. **b** Pairs $(\mathcal{K}_i, \mathcal{K}_j)$ of 5-node cliques are connected by a single link (top) to create a 24-clique ring network (bottom). The best modularity partition, which combines two neighboring cliques, has slightly larger modularity $Q_2 = 0.8712$ than the intuitive partition, which arranges individual cliques on their own, with modularity $Q_1 = 0.8674$. **c** The modularity landscape for this ring network displays a high-modularity plateau of degenerated solutions (inset). **d** A phenotype network for each possible three-input Boolean function, with each node representing one logic function. The functions that are most similar are linked together. The color of a node represents the modularity of the most cost-effective (wire-wise) feed-forward Boolean circuit for implementing that function. A light blue square **f** and a circle **h** highlight two important functions: "multiplexor" and "majority" respectively. Evolving a circuit while simultaneously minimizing the number of wires and implementing these target functions inevitably reduces the modularity of its predecessors (e $\rightarrow$ f and g $\rightarrow$ h). The "breakdown of modularity" refers to the transition from modular ancestors to less modular target circuits (bottom panels), as depicted in the phenotypic network (light blue arrows)

ing to reduce costs and enhance overall reliability [7]. Herbert Simon postulated that a modular system can adapt more rapidly than one that is not modular [86]. But whether modularity is a reflection of general evolutionary principles or is exclusive to certain functional features is still an open question. The prevalence of modular designs in both natural and human-made contexts suggests the existence of universal processes, while current heuristic methods for identifying modularity paint a different picture.

According to Kashtan and Alon [54], networks evolved under "modularly varying goals" (MVG) should also be modular. A network evolves under MVG when it adapts to changing objectives over time, with each goal having the same subproblems as the preceding one. An example is a logic circuit with multiple parts (modules), each implementing a sub-function needed to perform the overall objective. When objectives change, mutations that rewire these modules rapidly become permanent in the population to meet the new objective. Computational investigations have shown that MVG may greatly speed evolution (resulting in a modular network) as compared to evolution with fixed or fluctuating objectives (leading to a non-modular network). However, it is unclear how many biological settings vary in this modular fashion and if they change often enough to produce modular systems.

Modularity has also been suggested to evolve, not because it promotes evolvability in fluctuating environments, but as a by-product of selection that limits the number of links [25]. A more intriguing possibility is that mutation mechanisms, like gene duplication, could create a bias towards modular structures [92]. In the field of technology, makers employ previously accessible innovations as an integral aspect of their work, such as by replicating designs and reusing components acquired from others. Technology, like evolution, evolves mostly by tinkering [52], that is, by combining prior components in novel ways [6]. Even in the absence of selection, a basic network model based on duplication and rewiring can predict not only the empirical degree distribution but also the frequency of motifs in the real systems [112].

Agreement between data and generative models lead us to propose that motif distribution is equivalent to the concept of evolutionary spandrel discussed by Gould and Lewontin [91]. A popular biological argument is that structural features (such as modularity) have been selected for the advantages they provide. The logic goes as follows: certain features improve stability, improving the odds of survival; hence, these traits must be evolutionary adaptations. Some features, however, are little more than a 'spandrel,' as Gould and Lewontin call them [47, 59]. For example, we may observe ornamented structures in the midst of four branches on the ceiling of a cathedral (see Fig. 6a). Although these decorations look like important functional elements, they really reside in triangular spaces known in architecture as spandrels, which are an inevitable outcome of having two arches. Similarly, even in the absence of selection, the overabundance of certain subgraphs in networks may be an unavoidable result of tinkering processes (see Fig. 6b).

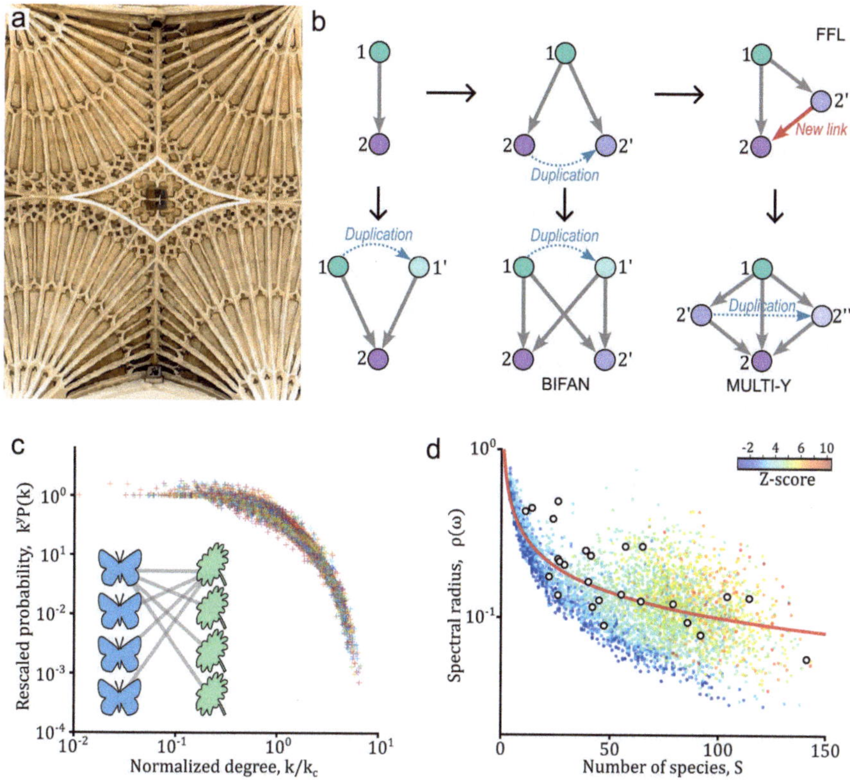

Fig. 6 Networks as evolutionary spandrels. Some network properties might be "spandrels," or byproducts or unintended consequences of tinkering rules, rather than the outcome of selection forces. **a** The ceiling of this cathedral shows decorated structures in the middle of four branches (surrounded by a white line). These decorations may be understood as functional essential pieces, although the space (an architectural spandrel) is just an unavoidable result of two arches. **b** We can generate several motifs by copying nodes (blue arrow) or adding new links (red arrow) starting from the simplest motif (top-left). This set of tinkering rules can generate three common motifs (FFL, BIFAN, and MULTI-Y) in directed networks, the **c** truncated power-law behaviour of mutualistic networks (inset) and the **d** relationship between the spectral radius ρ (a robust measure of nestedness, see [96]) and the number of species (S) (color nodes). Open circles correspond to a total of 25 real mutualistic networks [109]. The red line represents the random network prediction for nestedness, which is based on Wigner's semi-circle rule [4] and describes the average behavior but not the dispersion of real systems

3.4 Nestedness

A widespread and intriguing pattern in ecological networks is nestedness, which is characterized by the tendency of low-degree species to interact with a subset of highly connected species [13]. The canonical interpretation of nestedness in ecological networks is linked to the likelihood observing a particular pattern of interactions compared to a null model [121]. This analysis is traditionally carried out on topological matrices (meaning that they only account for the presence or absence of links). Such matrices are typically nested, but it has been pointed out [96], that a better characterization of these webs using link weights reveals that only a small fraction of them exhibit nestedness [100], suggesting that a truly meaningful assessment of nested patterns requires a weighted interaction matrix.

Edges in a nested network are organized in such a way that specialists interact with a subset of the species whom generalists interact with. This definition has been recently extended to quantitative networks using spectral graph theory [96, 109] (see Fig. 6d). Formally, nestedness is a systematic arrangement of non-zero entries in the adjacency matrix. This pattern can be clearly identified by computing the so-called overlap and declining fill (NODF[3]). For the bipartite graph $G = (P, M, E)$ with biadjacency matrix $B = [B_{ij}]$:

$$NODF(G) = \frac{1}{K}\left[\sum_{i,j=1}^{N_P}\left(\theta(q_i - q_j)\frac{\sum_{k=1}^{N_M} B_{ik}B_{jk}}{q_j}\right) + \sum_{k,l=1}^{N_M}\left(\theta(r_k - r_l)\frac{\sum_{i=1}^{N_P} B_{ik}B_{il}}{r_l}\right)\right]$$

where $K = [N_P(N_P - 1) + N_M(N_M - 1)]/200$ is a normalisation constant to ensure that $0 \leq NODF \leq 100$, $N_P = |||P||$, $N_M = ||M||$, θ is the Heaviside function with $\theta(0) = 0$, and q_i and r_j are the degrees of nodes $i \in P$ and $j \in M$, respectively. A high NODF value implies that certain species' interactions are a subset of other generalist species' interactions (and so the network demonstrates nestedness), whereas a low value indicates clustering (thus implying modular structures).

Following an adaptationist view of naturally evolved systems, it has been argued that the presence of nested patterns is a consequence of underlying selection processes that reduce competition relative to the benefits of facilitation, increasing biodiversity [12] and food web persistence [61]. Supporting evidence for this view are analytical approaches to generalized Lotka–Volterra equations with different functional responses. However, recent papers have challenged this view [100]. Instead, it has been argued that nestedness is likely to be a consequence (rather than a causative property) of biodiversity, in particular of the heterogeneous distributions of connections [29, 96] (see Fig. 6c). Furthermore, null models [109] have demonstrated that nestedness can be easily generated from generative processes; this suggests that tinkering models, which rely on basic copy and rewire processes, generate nested structures, thereby implying that nestedness is an evolutionary spandrel [47, 109]. Thus, validation of empirical nested patterns should take into account the predictions of tinkering models as a baseline to assess whether the observed patterns are meaningful, i.e. deviating from null expectations (see Fig. 6d).

4 Beyond Networks

Networks offer an efficient means of capturing both local and global structural features, serving as a valuable framework for investigating the dynamic properties of systems. However, their scope is limited to such pairwise interactions. In contrast, hypergraphs provide a solution to this limitation by allowing edges to connect any number of vertices (see Fig. 7b), rendering them exponentially more powerful and expressive in describing real-world interactions. Following previous notation, a hypergraph $H = (V, E)$ is a graph consisting of nodes (or vertices) $v_i \in V$ and and hyperlinks (or hyperedges) which can span any number of nodes $(v_i \ldots v_j) \in E$.

For instance, hypergraphs are indispensable when pathogens exhibit varied infection behaviors based on the presence or absence of multiple disease vectors [80] (see Fig. 7d), or in cases of symbiotic relationships involving more than two species [34, 67]. In other ecological systems, conventional depictions of predatory interactions, such as those in food webs, typically involve only two species. However, research has repeatedly demonstrated that many ecological interactions are of higher order [1]. This means that they can involve more than two species [44], incorporate external ecological factors that influence pairwise interactions [61, 116] (see Fig. 7e), or are driven by density-dependent behavior (such as when prey change their behaviour leading to their predators switching to other more suitable preys). In these cases, a network formalism may fall short in capturing the essential structural properties that drive emergent dynamical features. Hypergraphs become a necessary tool for making accurate predictions in systems with higher order interactions, due to their enhanced expressiveness and flexibility, rendering them highly valuable in complex systems research.

Given that the structure of ecological networks has been linked to patterns of extinction and robustness [17], how does a hypergraph perspective address these views? It has long been proposed that a modular network structure prevent the spread of perturbations (such as secondary extinctions) in ecological networks, with modules acting as firewalls against the propagation of extinction cascades [73, 98]. This suggests that modular patterns should be more resilient and, consequently, preserved in evolutionary processes. Conversely, mutualistic networks tend to exhibit a nested rather than modular pattern, which has also been associated with increased resilience and productivity in various ecological systems [12]. This assessment goes beyond mere conjecture; Sanders and colleagues demonstrated through field experiments with plant-insect communities that more complex food webs can indeed buffer against the effects of species loss [81].

Traditionally, modularity and nestedness were considered mutually exclusive [41, 96] and imbued ecological networks with different dynamical properties [12]. Various network models have emerged, attempting to account for their independent origins. However, recent research in hypergraphs has revealed that when modularity and nestedness inhabit distinct dimensions of high-order interaction graphs, or hypergraphs, they can coexist [116]. In this way, a projection of a tripartite hyper-

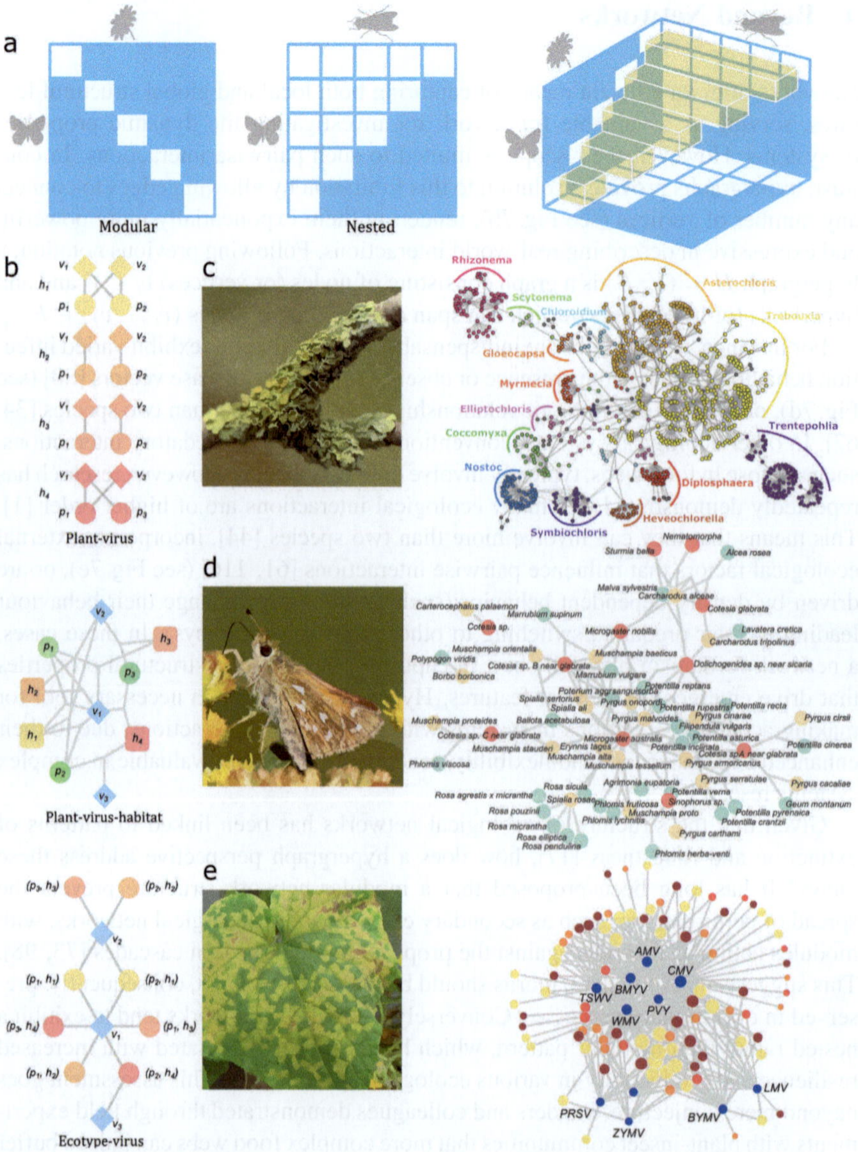

Fig. 7 Hypergraphs examples and bipartite projections. A tripartite hypergraph of host-parasites-habitat can be decomposed or projected into host-parasite or host-plant bipartite networks (**a**). These projections can display distinct structural patterns such as the host-plant network being modular (left) and the host-parasite being nested (right). Hypergraphs can also be reconstructed from 'partial' information, such as separate bipartite networks (**b**). Here, two bipartite networks are integrated into a hypergraph containing plants, viruses and biomes [116]. Many systems are amenable to a hypergraph representation: from lichen symbionts that contain many fungi-algal-bacterial relations [34] (**c**), plant-butterfly-parasite interactions [102] (**d**) and plant ecotype-parasite infection [116] (**e**)

graph into a network (by collapsing the adjacency matrix from 3 to 2 dimensions) can recover either structural pattern (see Fig. 7a).

Hypergraphs can also help researchers integrate different types of interactions in a single graph object. For instance, to better understand the compounding effect of human laws in biodiversity and biological interactions [37, 122]; the complex relationships between landscape, species and ecosystem services [40]; how anthropogenic climate change drives the reorganization of ecological networks [45]. With hypergraphs it also becomes possible to incorporate spatial or temporal dimensions to ecological networks [116], allowing us to address when structural patterns are driven by seasonal variation, or when interactions are guided by 'ecotypes' (see Fig. 7e, i.e. when species change their morphological features depending the biome they inhabit [51], thus altering the suite of interactions they partake in). However, although many of the structural metrics mentioned in this chapter also exist for hypergraphs [14], the field of empirical research in hypergraphs is still pretty much in its infancy, and further examples and mechanistic models will be necessary to bring them up to par to the ubiquity of network analysis.

5 Discussion

Biological complexity results from adaptations and shifts to novel features, taking place at various scales. Not all network characteristics, however, are the outcome of direct selection processes; rather, they can unexpectedly emerge from internal dynamics, akin to evolutionary spandrels. Scale-free degree distributions, motif frequencies, and even nestedness may have emerged as accidental outcomes of non-adaptive processes, such as tinkering rules. Simple growth models, however, are insufficient to explain how network complexity has evolved over time. Exogenous and endogenous mechanisms both contribute to the emergence of complexity.

Not all innovations lead to more complexity; the choice of which ones persist is ultimately based on environmental conditions and cost-benefit analyses. In technology, software engineers try to manage complexity by rationally subdividing a system into modules that interact in a neat and clear way [7]. External sources of complexity, like performance constraints, can entangle the original design, which soon becomes much less modular [105]. Software developers actively oppose complexity in this situation because unwanted complexity makes the system much more challenging to understand and maintain, often leading to entangled and monolithic systems [35]. Another source of complexity is the design principle that maximizes reuse of existing components. The merging of an existing system with an external one, e.g., a third-party library, results in complex co-evolutionary dynamics between different software projects [103].

We can use network morphospaces to characterise the vast landscape of network designs and how external forces and internal constraints drive structural diversity. These are representations of design spaces, where different networks are located according to some well defined metrics. Typically, the morphospace dimensions

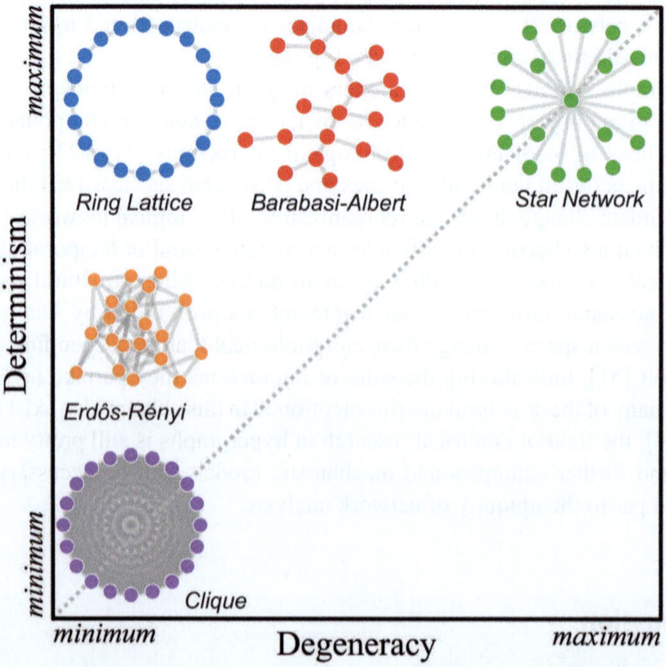

Fig. 8 Network morphospaces. Degeneracy $\langle H(W_i^{out})\rangle$ and determinism $H(\langle W_i^{out}\rangle)$ display a fundamental trade-off which confines networks to a specific domain of this design space [57]. Here, $H(X) = -\sum_n P(x_i)\log_2 P(x_i)$ represents the Shannon's entropy of a given distribution X, and the distribution $W_i^{out} = \{w_{ij}\}/\sum_j w_{ij}$ is the normalized set of weights from node i to every other node j. These weights can be understood as the distribution of probabilities a random walker might take when visiting node i. Regular structures (star, clique), lattices (ring) and typical networks produced by standard generative models (Erdős-Rényi and Barabási-Albert) are shown in this space. A forbidden region in the bottom right corner of the morphospace is uninhabited by any real network

are chosen because they reflect some deeper insight into the evolutionary process or constrains that limit what is possible in these systems, i.e. they often display trade-offs between them. One such example is the network degeneracy-determinism morphospace (see Fig. 8). This space quantifies the information contained in the topology of a network and can be used to signal the presence of higher informative scales. Here, determinism corresponds to the degree of determinism or certainty associated to a random walker traversing the network in every node, and the second dimension captures the system's degeneracy or entropy in weight distributions [57]. The features of this morphospace have implications for the dynamical properties of brain activity [50].

In this chapter we have explored a wealth of tools, metrics and morphospace representations that allow us to map and navigate complex systems. Understanding complexity not only gives us insights into the generative processes that underlie biological and technological systems, but also informs the software and biological engineers that want to manage the complexity of their designs. Stephen Hawking's prediction of a 'century of complexity' seems to be spot on with our current reality, which is distinguished by extraordinary technological advances contrasted against profound social and planetary challenges. In this age of complexity, our analysis of network architectures barely scratches the surface. Underlying there is a vast, undiscovered terrain, just like the bigger complexities awaiting discovery and comprehension in the world around us.

Acknowledgements This chapter is based on S. V.'s Introduction to Complex Networks course notes from the University of Valencia's Master of Computational Systems Biology programme. S.V. thanks Professor Juli Peretó's support to participate in the Master. S.V. & B.V. thanks Michael Stich for his kind invitation to participate in the 3rd Workshop of Nonlinear Dynamics in Biological Systems. S.V. is supported by the Spanish Ministry of Science and Innovation through the State Research Agency (AEI), grant PID2020-117822GB-I00 /AEI/10.13039/501100011033. B.V. and S.V. are supported through the 2020–2021 Biodiversa and Water JPI joint call under the BiodivRestore ERA-NET Cofund (GA N°101003777) project MPA4Sustainability with funding organizations: Innovation Fund Denmark (IFD), Agence Nationale de la Recherche (ANR), Fundação para a Ciência e a Tecnologia (FCT), Swedish Environmental Protection Agency (SEPA), and grant PCI2022-132936 funded by MCIN/AEI/10.13039/501100011033 and by the European Union NextGenerationEU/PRTR. B. V. was co-funded by the European Union (GA#101059915 - BIOcean5D). S.D-N. is supported by the Beatriu de Pinós postdoctoral programme, from the Office of the General Secretary of Research and Universities and the Ministry of Research and Univertisites (2019 BP 00206) and the support of the Marie Sklodowska-Curie COFUND (BP3 contract no. 801370) of the H2020 programme.

References

1. Abrams, P.A.: Arguments in favor of higher order interactions. Am. Nat. **121**, 887–891 (1983)
2. Alexander-Bloch, A.F., Gogtay, N., Meunier, D., Birn, R., Clasen, L., Lalonde, F., Lenroot, R., Giedd, J., Bullmore, E.T.: Disrupted modularity and local connectivity of brain functional networks in childhood-onset schizophrenia. Front. Syst. Neurosci. **4**, 147 (2010)
3. Almeida-Neto, M., Ulrich, W.: A straightforward computational approach for measuring nestedness using quantitative matrices. Environ. Model. Softw. **26**, 173–178 (2011)
4. Arnold, L.: On wigner's semicircle law for the eigenvalues of random matrices. Zeitschrift für Wahrscheinlichkeitstheorie und verwandte Gebiete **19**, 191–198 (1971)
5. Arthur, W.B.: Increasing Returns and Path Dependence in the Economy. University of Michigan Press (1994)
6. Arthur, W.B.: The Nature of Technology: What It Is and How It Evolves. Simon and Schuster (2009)
7. Baldwin, C.Y., Clark, K.B.: Design Rules: The Power of Modularity, vol. 1. MIT Press (2000)
8. Barabasi, A.L.: Network Science: The Scale-Free Property, pp. 1–15. Cambridge University Press (2016)
9. Barabási, A.L., Albert, R.: Emergence of scaling in random networks. Science **286**, 509–512 (1999)
10. Barthelemy, M.: Morphogenesis of Spatial Networks. Springer (2018)
11. Basalla, G.: The Evolution of Technology. Cambridge University Press (1988)
12. Bascompte, J., Jordano, P.: Plant-animal mutualistic networks: the architecture of biodiversity. Annu. Rev. Ecol. Evol. Syst. **38**, 567–593 (2007)

13. Bascompte, J., Jordano, P., Melián, C.J., Olesen, J.M.: The nested assembly of plant-animal mutualistic networks. Proc. Natl. Acad. Sci. **100**, 9383–9387 (2003)
14. Battiston, F., Cencetti, G., Iacopini, I., Latora, V., Lucas, M., Patania, A., Young, J.G., Petri, G.: Networks beyond pairwise interactions: structure and dynamics. Phys. Rep. **874**, 1–92 (2020)
15. Batty, M., Longley, P.A.: Fractal Cities: A Geometry of Form and Function. Academic Press (1994)
16. Bonner, J.T.: The Evolution of Complexity by Means of Natural Selection. Princeton University Press (1988)
17. Bornholdt, S., Sneppen, K.: Robustness as an evolutionary principle. Proc. R. Soc. Lond. Ser. B Biol. Sci. **267**, 2281–2286 (2000)
18. Brandes, U., Delling, D., Gaertler, M., Görke, R., Hoefer, M., Nikoloski, Z., Wagner, D.: Maximizing modularity is hard (2006). arXiv preprint physics/0608255
19. Brunet, T., King, N.: The origin of animal multicellularity and cell differentiation. Dev. Cell **43**, 124–140 (2017)
20. Burt, R.S.: Structural holes and good ideas. Am. J. Sociol. **110**, 349–399 (2004)
21. Callaway, D., Newman, M., Strogatz, S., Watts, D.: Exact solution of percolation on random graphs with arbitrary degree distributions. Phys. Rev. Lett. **85**, 5468–5471 (2000)
22. Carbonell-Ballestero, M., Duran-Nebreda, S., Montanez, R., Solé, R., Macía, J., Rodríguez-Caso, C.: A bottom-up characterization of transfer functions for synthetic biology designs: lessons from enzymology. Nucleic Acids Res. **42**, 14060–14069 (2014)
23. Carbonell-Ballestero, M., Garcia-Ramallo, E., Montañez, R., Rodriguez-Caso, C., Macía, J.: Dealing with the genetic load in bacterial synthetic biology circuits: convergences with the Ohm's law. Nucleic Acids Res. **44**, 496–507 (2016)
24. Cartwright, T.J.: Planning and chaos theory. J. Am. Plan. Assoc. **57**, 44–56 (1991)
25. Clune, J., Mouret, J.B., Lipson, H.: The evolutionary origins of modularity. Proc. R. Soc. B Biol. Sci. **280**, 20122863 (2013)
26. Corominas-Murtra, B., Valverde, S., Solé, R.: The ontogeny of scale-free syntax networks: phase transitions in early language acquisition. Adv. Complex Syst. **12**, 371–392 (2009)
27. David, A.S.: Dysmodularity: a neurocognitive model for schizophrenia. Schizophrenia Bull. **20**, 249–255 (1994)
28. Dey, A.K., Gel, Y.R., Poor, H.V.: What network motifs tell us about resilience and reliability of complex networks. Proc. Natl. Acad. Sci. **116**, 19368–19373 (2019)
29. Dunne, J.A., Williams, R.J., Martinez, N.D.: Network structure and biodiversity loss in food webs: robustness increases with connectance. Ecol. Lett. **5**, 558–567 (2002)
30. Duran-Nebreda, S., Bassel, G.W.: Bridging scales in plant biology using network science. Trends Plant Sci. **22**, 1001–1003 (2017)
31. Duran-Nebreda, S., Bonforti, A., Montanez, R., Valverde, S., Solé, R.: Emergence of proto-organisms from bistable stochastic differentiation and adhesion. J. R. Soc. Interface **13**, 20160108 (2016)
32. Duran-Nebreda, S., O'Brien, M.J., Bentley, R.A., Valverde, S.: Dilution of expertise in the rise and fall of collective innovation. Hum. Soc. Sci. Commun. **9**, 1–10 (2022)
33. Duran-Nebreda, S., Pla, J., Vidiella, B., Pinero, J., Conde-Pueyo, N., Solé, R.: Synthetic lateral inhibition in periodic pattern forming microbial colonies. ACS Synth. Biol. **10**, 277–285 (2021)
34. Duran-Nebreda, S., Valverde, S.: Composition, structure and robustness of lichen guilds. Sci. Rep. **13**, 3295 (2023)
35. Eick, S.G., Graves, T.L., Karr, A.F., Marron, J.S., Mockus, A.: Does code decay? Assessing the evidence from change management data. IEEE Trans. Softw. Eng. **27**, 1–12 (2001)
36. Elowitz, M.B., Leibler, S.: A synthetic oscillatory network of transcriptional regulators. Nature **403**, 335–338 (2000)
37. Epstein, Y., López-Bao, J.V., Chapron, G.: A legal-ecological understanding of favorable conservation status for species in Europe. Conserv. Lett. **9**, 81–88 (2016)

38. Érdi, P., Makovi, K., Somogyvári, Z., Strandburg, K., Tobochnik, J., Volf, P., Zalányi, L.: Prediction of emerging technologies based on analysis of the us patent citation network. Scientometrics **95**, 225–242 (2013)
39. Erdős, P., Rényi, A.: On the evolution of random graphs. Publ. Math. Inst. Hung. Acad. Sci. **5**, 17–60 (1960)
40. Felipe-Lucia, M.R., Soliveres, S., Penone, C., Fischer, M., Ammer, C., Boch, S., Boeddinghaus, R.S., Bonkowski, M., Buscot, F., Fiore-Donno, A.M., et al.: Land-use intensity alters networks between biodiversity, ecosystem functions, and services. Proc. Natl. Acad. Sci. **117**, 28140–28149 (2020)
41. Fortuna, M.A., Stouffer, D.B., Olesen, J.M., Jordano, P., Mouillot, D., Krasnov, B.R., Poulin, R., Bascompte, J.: Nestedness versus modularity in ecological networks: two sides of the same coin? J. Anim. Ecol. 811–817 (2010)
42. Fortunato, S., Barthelemy, M.: Resolution limit in community detection. Proc. Natl. Acad. Sci. **104**, 36–41 (2007)
43. Godwin, D., Barry, R.L., Marois, R.: Breakdown of the brain's functional network modularity with awareness. Proc. Natl. Acad. Sci. **112**, 3799–3804 (2015)
44. Golubski, A.J., Westlund, E.E., Vandermeer, J., Pascual, M.: Ecological networks over the edge: hypergraph trait-mediated indirect interaction (TMII) structure. Trends Ecol. Evol. **31**, 344–354 (2016)
45. Gómez-Ruiz, E.P., Lacher, T.E., Jr.: Climate change, range shifts, and the disruption of a pollinator-plant complex. Sci. Rep. **9**, 14048 (2019)
46. Good, B.H., De Montjoye, Y.A., Clauset, A.: Performance of modularity maximization in practical contexts. Phys. Rev. E **81**, 046106 (2010)
47. Gould, S.J., Lewontin, R.C.: The spandrels of San Marco and the Panglossian paradigm: a critique of the adaptationist programme. Proc. R. Soc. Lond. B **205**, 581–598 (1979)
48. Gould, S.J., Vrba, E.S.: Exaptation—a missing term in the science of form. Paleobiology **8**, 4–15 (1982)
49. Handcock, M.S., Raftery, A.E., Tantrum, J.M.: Model-based clustering for social networks. J. R. Stat. Soc. Ser. A (Stat. Soc.) **170**, 301–354 (2007)
50. Hoel, E.P., Albantakis, L., Marshall, W., Tononi, G.: Can the macro beat the micro? Integrated information across spatiotemporal scales. Neurosci. Conscious. **2016**, niw012 (2016)
51. Jackson, M.D., Xu, H., Duran-Nebreda, S., Stamm, P., Bassel, G.W.: Topological analysis of multicellular complexity in the plant hypocotyl. Elife **6**, e26023 (2017)
52. Jacob, F.: Evolution and tinkering. Science **196**, 1161–1166 (1977)
53. Jonas, K., Chen, Y.E., Laub, M.T.: Modularity of the bacterial cell cycle enables independent spatial and temporal control of dna replication. Curr. Biol. **21**, 1092–1101 (2011)
54. Kashtan, N., Alon, U.: Spontaneous evolution of modularity and network motifs. Proc. Natl. Acad. Sci. **102**, 13773–13778 (2005)
55. Kay, A.C.: The early history of smalltalk. ACM SIGPLAN Notices **28**, 69–95 (1993)
56. Kesten, H., et al.: The critical probability of bond percolation on the square lattice equals 1/2. Commun. Math. Phys. **74**, 41–59 (1980)
57. Klein, B., Hoel, E.: The emergence of informative higher scales in complex networks. Complexity **2020**, 1–12 (2020)
58. Knabe, J.F., Nehaniv, C.L., Schilstra, M.J.: Do motifs reflect evolved function?—No convergent evolution of genetic regulatory network subgraph topologies. Biosystems **94**, 68–74 (2008)
59. Koonin, E.V.: Splendor and misery of adaptation, or the importance of neutral null for understanding evolution. BMC Biol. **14**, 1–8 (2016)
60. Lacasa, L., Luque, B., Ballesteros, F., Luque, J., Nuno, J.C.: From time series to complex networks: the visibility graph. Proc. Natl. Acad. Sci. **105**, 4972–4975 (2008)
61. Levine, J.M., Bascompte, J., Adler, P.B., Allesina, S.: Beyond pairwise mechanisms of species coexistence in complex communities. Nature **546**, 56–64 (2017)
62. Macia, J., Vidiella, B., Solé, R.V.: Synthetic associative learning in engineered multicellular consortia. J. R. Soc. Interface **14**, 20170158 (2017)

63. Madsen, C., Goñi Moreno, A., P, U., Palchick, Z., Roehner, N., Atallah, C., Bartley, B., Choi, K., Cox, R.S., Gorochowski, T., et al.: Synthetic biology open language (SBOL) version 2.3. J. Integr. Bioinform. **16**, 20190025 (2019)
64. Maslov, S., Sneppen, K.: Specificity and stability in topology of protein networks. Science **296**, 910–913 (2002)
65. Mesoudi, A.: Cultural selection and biased transformation: two dynamics of cultural evolution. Philos. Trans. R. Soc. B **376**, 20200053 (2021)
66. Milo, R., Shen-Orr, S., Itzkovitz, S., Kashtan, N., Chklovskii, D., Alon, U.: Network motifs: simple building blocks of complex networks. Science **298**, 824–827 (2002)
67. Miransari, M.: Interactions between arbuscular mycorrhizal fungi and soil bacteria. Appl. Microbiol. Biotechnol. **89**, 917–930 (2011)
68. Mirzasoleiman, B., Jalili, M.: Failure tolerance of motif structure in biological networks. PLoS One **6**, e20512 (2011)
69. Newberry, M.G., Plotkin, J.B.: Measuring frequency-dependent selection in culture. Nat. Hum. Behav. **6**, 1048–1055 (2022)
70. Newman, M.: Networks. Oxford University Press (2018)
71. Newman, M.E., Girvan, M.: Finding and evaluating community structure in networks. Phys. Rev. E **69**, 026113 (2004)
72. O'Brien, M.J., Lyman, R.L.: Applying Evolutionary Archaeology: A Systematic Approach. Springer Science & Business Media (2000)
73. Olesen, J.M., Bascompte, J., Dupont, Y.L., Jordano, P.: The modularity of pollination networks. Proc. Natl. Acad. Sci. **104**, 19891–19896 (2007)
74. Otte, E., Rousseau, R.: Social network analysis: a powerful strategy, also for the information sciences. J. Inf. Sci. **28**, 441–453 (2002)
75. Perna, A., Jost, C., Couturier, E., Valverde, S., Douady, S., Theraulaz, G.: The structure of gallery networks in the nests of termite Cubitermes spp. revealed by x-ray tomography. Naturwissenschaften **95**, 877–884 (2008)
76. Poisot, T., Gravel, D.: When is an ecological network complex? Connectance drives degree distribution and emerging network properties. PeerJ **2**, e251 (2014)
77. Radicchi, F., Castellano, C.: Beyond the locally treelike approximation for percolation on real networks. Phys. Rev. E **93**, 030302 (2016)
78. Ratcliff, W.C., Denison, R.F., Borrello, M., Travisano, M.: Experimental evolution of multicellularity. Proc. Natl. Acad. Sci. **109**, 1595–1600 (2012)
79. Rives, A.W., Galitski, T.: Modular organization of cellular networks. Proc. Natl. Acad. Sci. **100**, 1128–1133 (2003)
80. Rizzoli, A., Tagliapietra, V., Cagnacci, F., Marini, G., Arnoldi, D., Rosso, F., Rosà, R.: Parasites and wildlife in a changing world: the vector-host-pathogen interaction as a learning case. Int. J. Parasitol. Parasites Wildlife **9**, 394–401 (2019)
81. Sanders, D., Thébault, E., Kehoe, R., Frank van Veen, F.: Trophic redundancy reduces vulnerability to extinction cascades. Proc. Natl. Acad. Sci. **115**, 2419–2424 (2018)
82. Schaerli, Y., Munteanu, A., Gili, M., Cotterell, J., Sharpe, J., Isalan, M.: A unified design space of synthetic stripe-forming networks. Nat. Commun. **5**, 4905 (2014)
83. Schlosser, G., Wagner, G.P.: Modularity in Development and Evolution. University of Chicago Press (2004)
84. Schülke, C., Ricci-Tersenghi, F.: Multiple phases in modularity-based community detection. Phys. Rev. E **92**, 042804 (2015)
85. Simon, H.A.: On a class of skew distribution functions. Biometrika **42**, 425–440 (1955)
86. Simon, H.A.: The architecture of complexity. Proc. Am. Philos. Soc. **106**, 467–482 (1962)
87. Solé, R.: Phase Transitions, vol. 3. Princeton University Press (2011)
88. Solé, R., Amor, D.R., Duran-Nebreda, S., Conde-Pueyo, N., Carbonell-Ballestero, M., Montañez, R.: Synthetic collective intelligence. Biosystems **148**, 47–61 (2016)
89. Solé, R., Ollé-Vila, A., Vidiella, B., Duran-Nebreda, S., Conde-Pueyo, N.: The road to synthetic multicellularity. Curr. Opin. Syst. Biol. **7**, 60–67 (2018)

90. Solé, R.V., Valverde, S.: Information theory of complex networks: on evolution and architectural constraints. In: Complex Networks, pp. 189–207. Springer (2004)
91. Solé, R.V., Valverde, S.: Are network motifs the spandrels of cellular complexity? Trends Ecol. Evol. **21**, 419–422 (2006)
92. Solé, R.V., Valverde, S.: Spontaneous emergence of modularity in cellular networks. J. R. Soc. Interface **5**, 129–133 (2008)
93. Solé, R.V., Valverde, S., Casals, M.R., Kauffman, S.A., Farmer, D., Eldredge, N.: The evolutionary ecology of technological innovations. Complexity **18**, 15–27 (2013)
94. de Solla Price, D.: Networks of scientific papers. Science (New York, NY) **149**, 510–515 (1965)
95. Sporns, O., Kötter, R.: Motifs in brain networks. PLoS Biol. **2**, e369 (2004)
96. Staniczenko, P.P., Kopp, J.C., Allesina, S.: The ghost of nestedness in ecological networks. Nat. Commun. **4**, 1–6 (2013)
97. Steadman, P.: The Evolution of Designs: Biological Analogy in Architecture and the Applied Arts. Routledge (2008)
98. Stouffer, D.B., Bascompte, J.: Compartmentalization increases food-web persistence. Proc. Natl. Acad. Sci. **108**, 3648–3652 (2011)
99. Szathmáry, E.: Toward major evolutionary transitions theory 2.0. Proc. Natl. Acad. Sci. **112**, 10104–10111 (2015)
100. Timóteo, S., Albrecht, J., Rumeu, B., Norte, A.C., Traveset, A., Frost, C.M., Marchante, E., López-Núñez, F.A., Peralta, G., Memmott, J., et al.: Tripartite networks show that keystone species can multitask. Func. Ecol. (2022)
101. Tononi, G., Sporns, O., Edelman, G.M.: Measures of degeneracy and redundancy in biological networks. Proc. Natl. Acad. Sci. **96**, 3257–3262 (1999)
102. Toro-Delgado, E., Hernández-Roldán, J., Dincă, V., Vicente, J.C., Shaw, M.R., Quicke, D.L., Vodă, R., Albrecht, M., Fernández-Triana, J., Vidiella, B., et al.: Butterfly-parasitoid-hostplant interactions in Western Palaearctic hesperiidae: a DNA barcoding reference library. Zool. J. Linnean Soc. **196**, 757–774 (2022)
103. Valverde, S.: Crossover from endogenous to exogenous activity in open-source software development. Europhys. Lett. **77**, 20002 (2007)
104. Valverde, S.: Major transitions in information technology. Philos. Trans. R. Soc. B Biol. Sci. **371**, 20150450 (2016)
105. Valverde, S.: Breakdown of modularity in complex networks. Front. Physiol. **8**, 497 (2017)
106. Valverde, S.: The long and winding road: accidents and tinkering in software standardization. Metode Sci. Stud. J. 91–97 (2021). https://revistas.uv.es/index.php/Metode/article/view/16112; https://doi.org/10.7203/metode.11.16112
107. Valverde, S., Ferrer-i Cancho, R., Solé, R.V.: Scale-free networks from optimal design. Europhys. Lett. **60**, 512 (2002)
108. Valverde, S., Corominas-Murtra, B., Perna, A., Kuntz, P., Theraulaz, G., Solé, R.V.: Percolation in insect nest networks: evidence for optimal wiring. Phys. Rev. E **79**, 066106 (2009)
109. Valverde, S., Piñero, J., Corominas-Murtra, B., Montoya, J., Joppa, L., Solé, R.: The architecture of mutualistic networks as an evolutionary spandrel. Nat. Ecol. Evol. **2**, 94–99 (2018)
110. Valverde, S., Solé, R.: Logarithmic growth dynamics in software networks. Europhys. Lett. **72**, 858 (2005)
111. Valverde, S., Solé, R.V.: Internet's critical path horizon. Eur. Phys. J. B **38**, 245–252 (2004)
112. Valverde, S., Solé, R.V.: Network motifs in computational graphs: a case study in software architecture. Phys. Rev. E **72**, 026107 (2005)
113. Valverde, S., Solé, R.V.: Hierarchical small-worlds in software architecture. In: Teo, K.L., Liu, X. (eds.) Dynamics of Continuous Discrete Impulsive Systems Series B: Applications and Algorithms (Special Issue on Software Engineering and Complex Networks, vol. 14, pp. 1–11 (2007)
114. Valverde, S., Solé, R.V.: Networks and the city. Arch. Des. **83**, 112–119 (2013)
115. Valverde, S., Solé, R.V., Bedau, M.A., Packard, N.: Topology and evolution of technology innovation networks. Phys. Rev. E **76**, 056118 (2007)

116. Valverde, S., Vidiella, B., Montañez, R., Fraile, A., Sacristán, S., García-Arenal, F.: Coexistence of nestedness and modularity in host-pathogen infection networks. Nat. Ecol. Evol. **4**, 568–577 (2020)
117. Vidiella, B., Guillamon, A., Sardanyés, J., Maull, V., Pla, J., Conde, N., Solé, R.: Engineering self-organized criticality in living cells. Nat. Commun. **12**, 4415 (2021)
118. Watts, D.J., Strogatz, S.H.: Collective dynamics of 'small-world' networks. Nature **393**, 440–442 (1998)
119. Weinstock, M.: The Architecture of Emergence: The Evolution of Form in Nature and Civilisation. Wiley (2010)
120. Weinstock, M.: The metabolism of the city: the mathematics of networks and urban surfaces. Arch. Des. **81**, 102–107 (2011)
121. Weitz, J.S., Poisot, T., Meyer, J.R., Flores, C.O., Valverde, S., Sullivan, M.B., Hochberg, M.E.: Phage-bacteria infection networks. Trends Microbiol. **21**, 82–91 (2013)
122. Wells, E., Wilkinson, M., Wood, P., Scanlan, C.: The use of macroalgal species richness and composition on intertidal rocky seashores in the assessment of ecological quality under the European water framework directive. Mar. Pollut. Bull. **55**, 151–161 (2007)
123. Wilson, E.O., Hölldobler, B.: The Superorganism: The Beauty, Elegance, and Strangeness of Insect Societies. W. W Norton & Company, New York (2009)
124. Woodger, J.H., Tarski, A., Floyd, W.F.: The axiomatic method in biology. J. Unified Sci. (Erkenntnis) **8** (1940)
125. Yule, G.U.: II.—A mathematical theory of evolution, based on the conclusions of Dr. JC Willis, FR S. Philos. Trans. R. Soc. Lond. Ser. B Containing Papers Biol. Char. **213**, 21–87 (1925)

Modelling the Formation of the Granuloma Inside Secondary Lobule of the Lungs

Sergio Alonso, Júlia Cabrera, Martí Català, and Clara Prats

Abstract Tuberculosis is an important infectious disease caused by a bacterium and spread all over the world. Although some individuals develop a severe active tuberculosis infection (ATB), most of the individuals evolve into the asymptomatic Latent Tuberculosis Infection (LTBI), through the formation and control of granuloma inside a pulmonary secondary lobule. In such case, the growth of the lesion is restricted inside one of the secondary lobules. There are around 10^3 secondary lobules in a human lung. The development of the lesion inside the lung can be modeled by a set of reaction-diffusion equations involving different types of cells. However, here a reduced model has been developed to study the evolution of an initial infection until it is encapsulated, both in 2D and in 3D. For the 2D case, a sensitivity analysis has been performed to study the influence on the model's parameters to the final radius of the granuloma and the encapsulation time. For the 3D case, unstable and stable approaches have been considered and their encapsulation processes compared. The model was first implemented in a cube and afterwards extended to an irregular shaped secondary lobule. In the near future, we plan to apply such model to a virtual lung to first reproduce the results of a lung-scale agent-based model that fits the experimental results of the tuberculosis spreading along the whole lungs, and next to be able to understand the process involved in the differences between the active and the latent versions of the tuberculosis infections.

S. Alonso (✉) · J. Cabrera · C. Prats
Department of Physics, Universitat Politècnica de Catalunya, Barcelona, Spain
e-mail: s.alonso@upc.edu

J. Cabrera
e-mail: julia.cabrera@estudiantat.upc.edu

C. Prats
e-mail: clara.prats@upc.edu

M. Català
University of Oxford, Oxford, UK
e-mail: marti.catalasabate@ndorms.ox.ac.uk

© The Author(s), under exclusive license to Springer Nature Switzerland AG 2025

M. Stich and J. Carballido-Landeira (eds.), *Nonlinear Dynamics for Biological Systems*,
SEMA SIMAI Springer Series 40,
https://doi.org/10.1007/978-3-031-99044-1_7

1 Introduction

Tuberculosis is caused by the *Mycobacterium tuberculosis* (*Mtb*), and it is one of the world's most spread infectious disease. The World Health Organization [16] estimates that nowadays more than a quarter of the world's population is infected. However, most of the infected people have no symptoms and cannot infect other people because they undergo a Latent Tuberculosis Infection (LTBI). The mechanisms that lead to around 10% of these people to eventually develop active tuberculosis (ATB) are still unclear.

The natural history of tuberculosis (i.e., the dynamics of the tuberculosis infection) is complex and many parts of the process of infection can be studied separately: from the inflammatory response in one single alveolus to the spatial distribution of the lesions inside the lungs after several weeks of infection [2], see Fig. 1. The initial TB infection mainly occurs via air drops of a person with a pulmonary ATB that are inhaled by the susceptible host. When *Mtb* arrive to the alveoli, alveolar macrophages (AM) try to control them. Alveolar macrophages phagocyte the bacilli but, although this process may be effective to eliminate the majority of inhaled agents, *Mtb* is able to proliferate inside the AM until it is destructed, see Fig. 1a. After that, intracellular bacilli become extracellular and are again phagocyted by another AM. This cycle continues and the accumulated amount of dead AM and bacilli trigger the inflammatory response. This response consists of the entrance of uninfected macrophages and neutrophils into the infected alveolus, see Fig. 1b, which has important effects on the formation of the granuloma [7]. In general, these mechanisms are not always able to control the infection and a specific immune response is necessary. T-cells arrive to the infected alveoli and attach to AM, which, after that, are able to eventually eliminate the bacilli, see Fig. 1c. Unfortunately the AM activated by the T-cells are usually not able to stop the growing of the lesion and the growth is finally stopped by the encapsulation by the fibroblast, see Fig. 1d, which consists in building a wall of fibroblasts and collagen surrounding the lesion.

A human lung is divided in around 1000 secondary lobules, see Fig. 2a. The secondary pulmonary lobule is a physiologic unit of lung supplied by around four terminal bronchioles and separated from other lobules by fibrous septa, see Fig. 2b. This structure, which forms an irregular polyhedron measuring 1–3 cm in diameter, contains a variable number of alveoli sacs. All the elements taking part in the processes mentioned above may occupy several alveoli, being able to escape from the primary alveolus. Each secondary lobule is surrounded by pulmonary connective tissue forming the septa, responsible for the process of encapsulation of the granuloma. Connective tissue contains many types of cells, in particular fibroblasts, which are able to encapsulate the lesion, by the segregation of collagen, and stop it from growing. If the infection is controlled by the immune system and fibroblasts are able to efficiently perform the encapsulation, the infected individual is left in a state of LTBI, which means that the lesion is no more evolving and the individual does not have symptoms and cannot infect other people. The encapsulation process is complementary to the immune response.

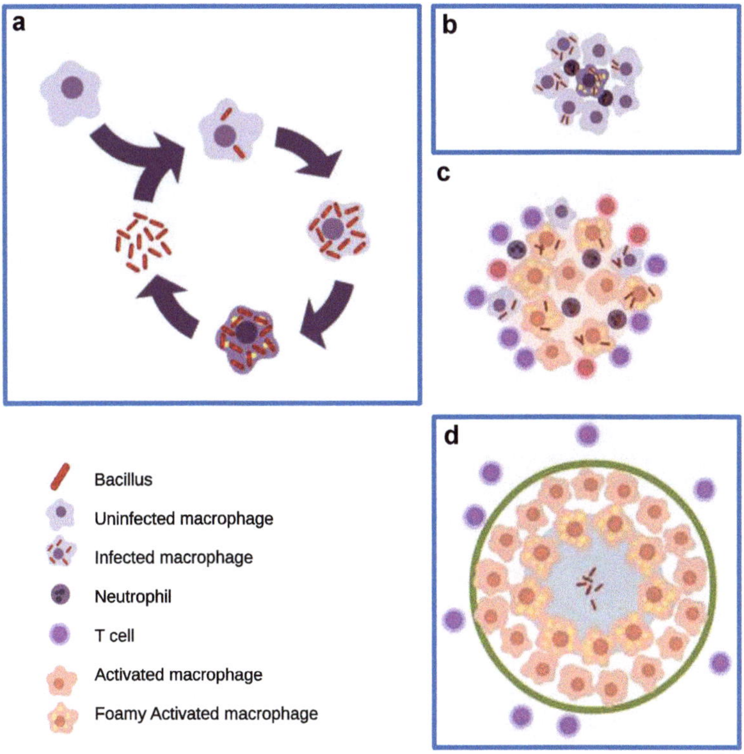

Fig. 1 Sketch of the dynamics of the initial infection, the interaction between bacilli and the alveolar macrophage (**a**), inflammatory response: triggered by the presence of extracellular bacilli (**b**), immune response: infected macrophages trigger the mechanisms that activate this response (**c**), and encapsulation: led by fibroblasts, it takes place when a lesion makes contact with a pulmonary membrane (**d**). We consider 5 types of cells in this dynamics: bacilli (extracellular or intracellular), macrophages (uninfected, infected, activated-foamy) neutrophils, T cells and fibroblasts. Figure adapted from [6]

Encapsulation is an efficient process depending on the size and bacillary load of the inflammation [15]. Normally, lesions which have a small bacillary load at the beginning can be encapsulated, leaving the patient in a LTBI state with a granuloma inside the secondary lobule. However, in the case of more severe lesions, bacillary load when first infected is high, making the inflammatory and immune response not able to stop the exponential growth of bacilli giving rise to ATB.

Different mathematical modelling strategies have been considered to model the processes involved in the dynamics of the tuberculosis infection [4]. In particular we focus on the formation of granulomas in the secondary lobule. Here, we address two models which have previously considered similar approaches

First, an *Agent Based Model (ABM)*, see Ref. [1] for a general review of the method. We have previously employed such type of methods for the study of infection

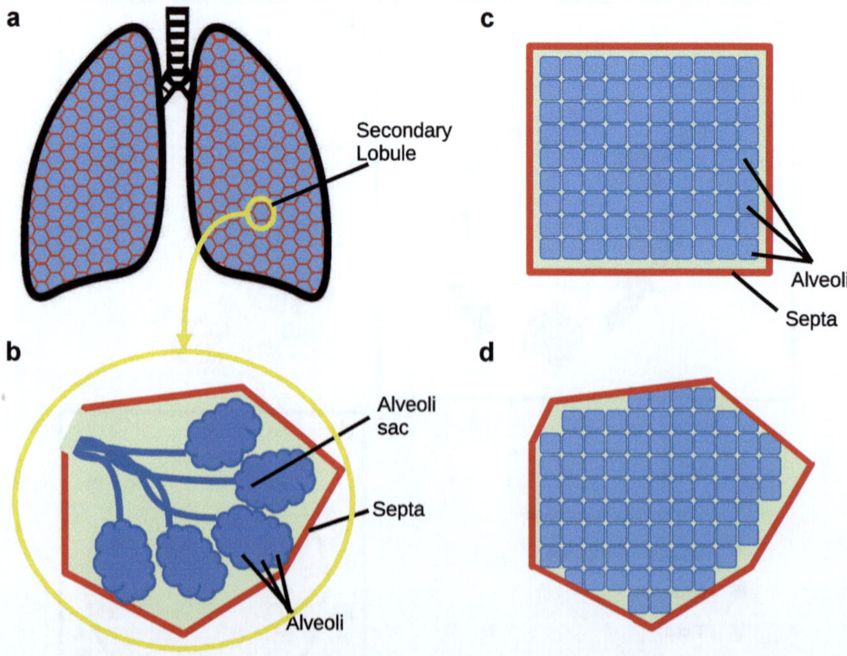

Fig. 2 Sketch of the division of the lungs in secondary lobules (**a**), sketch of the secondary lobule (**b**) and representation of the space on the regular (**c**) and non-regular (**d**) model represented in two dimensions

diseases [3]. For the modeling of the dynamics of the tuberculosis infection, we have also employed an ABM called *Bubble model* that aims to reproduce the dynamics of tuberculosis lesions at the level of the lung and that successfully reproduce observed experimental results in mice [12] and minipigs [5]. This model does not take into account such detailed processes of the evolution and rather focuses on the initial and final configuration of lesions all over the lung, how new lesions proliferate and how lesions merge if they are close enough. It was initially built to reproduce some experimental results regarding the evolution of tuberculosis lesions in female mice [10]. The relevant information extracted from the experiments were the initial and final number and distribution of tuberculosis lesions among all the lungs, as well as the evolution of their areas and the area distribution. This model considers every TB lesion as an individual element with some properties that are time dependent apart from age, such as their size. The model does not take into account that a lesion is actually formed by a huge number of *Mtb* and other cells, it does take into account the inflammatory and immune responses and how it affects to the evolution of the lesion since the moment that the lung is first infected.

Second, a *reaction-diffusion model*, see Ref. [8] for a general review of this method. We have previously employed such type of methods for the study of infection diseases [11]. For the modeling of the tuberculosis infection, we have also employed a reaction-diffusion model that takes into account all the interactions among the elements of the dynamics of TB and reproduces the evolution of the infection from the alveolus to the secondary lobule [6]. It was built to reproduce the dynamics of tuberculosis from the alveolus level to the secondary lobule level. The model considers all the different biological elements that take part in the evolution of the lesion: bacilli (*Mtb*), macrophages, neutrophils, T cells and fibroblasts. Considering the possible interaction among these elements, bacilli and macrophages lead to 5 types of cells: extracellular and intracellular bacilli, uninfected, infected or activated macrophages. The evolution of the initial infection is divided into 4 parts: (1) the initial infection, where bacilli infect macrophages, see Fig. 1a, and a considerable amount of dead macrophages is accumulated; (2) the inflammatory response, when there is a massive entrance of neutrophils and uninfected macrophages; this inflammatory response is not able to stop the evolution, see Fig. 1b, and, therefore, (3) immune response is triggered, when T cells activate macrophages giving them the ability to kill intracellular bacilli, see Fig. 1c; and (4) the encapsulation of the lesion carried out by the fibroblasts located in the connective tissue that surrounds the secondary lobule, see Fig. 1d. However, such model is too complex to be expanded to the lung scale, as it depends on too many parameters and that would be computationally demanding.

Here, we take an intermediate approach. The goal is to develop a reduced reaction-diffusion model that simplifies the evolution of the original reaction-diffusion model by reducing the number of parameters. In the future, we want this model to reproduce the obtained results from the Agent-based Model in the simulations inside virtual lungs. Two versions are considered: first, a single alveolus is modelled and the first three steps of the evolution can be studied in detail. The second part considers the first steps of the evolution at the level of the secondary lobule, and the process of encapsulation is more easily studied.

2 Model

The aim of the model is to study the formation of granulomas in the secondary lobule as a result of the process of encapsulation. It is based on a set of reaction-diffusion equations that reproduces the evolution of the lesion at the level of the secondary lobule, see Fig. 2a for the sketch of the division of the lungs in secondary lobule and Fig. 2b for a sketch of the secondary lobule. This set of alveoli is redistributed into a regular grid of alveoli inside a two or three dimensional space surrounded by the septa. This space can be reorganized in a regular, see Fig. 2c or in a non-regular, see Fig. 2d, network. The lesion formation is the result of processes on different scales that were not specifically modeled. The model focuses into reproducing two main processes: the initial lesion exponential growth, and the posterior encapsulation processes that stops lesions growth and stabilize their size. Encapsulation is triggered by fibroblast chemokinesis and the secretion of collagen.

The main mechanisms of the model that are taken into account are: (1) lesions grow in each alveolus and diffuse to neighboring spaces that have not been infected nor occupied by fibroblasts; (2) fibroblasts can detect the gradient of the lesion and start diffusing accordingly, maintaining regular flux from the borders of the secondary lobule; and (3) the fibroblasts and the segregated collagen around the lesion cannot occupy the same space than the lesion and, therefore, the encirclement prevents the lesion from growing.

The advantages of not taking into account all the details of the processes that bring the lesion to reach the secondary lobule, is that the number of parameters of the model is drastically reduced in comparison with the original reaction-diffusion model previously considered [6].

2.1 Single Alveolus Model

The lesion consists of the aggregation of bacilli, macrophages and necrotic material. From the initial infection the inflammatory response causes the lesion to grow exponentially $c = c_0 e^{kt}$, which is the solution of the equation:

$$\frac{\partial c}{\partial t} = kc \tag{1}$$

where c represents the fraction of volume of an alveolus occupied by the lesion, and the parameter k is the initial growth rate of the lesion. It means that the rate of reproduction is proportional to the actual fraction of volume occupied by bacilli. However there is a maximum for this fraction, and therefore the growth saturates at $c = 1$:

$$\frac{\partial c}{\partial t} = kc(1 - c) \tag{2}$$

The stationary condition permits to calculate the fixed points. As shown in Fig. 3a, the first stability condition corresponding to $c = 0$ is an unstable fixed point, and the second corresponding to $c = 1$ is stable. This fact makes the model unstable, because any perturbation will make the system evolve irreversibly until $c = 1$.

We can consider a second model where the fixed point at $c = 0$ is stable by adding a factor $(c - a)$, where a is a particular threshold between 0 and 1, see Fig. 3b. Below this value the fraction decreases to zero and above this value grows:

$$\frac{\partial c}{\partial t} = kc(1 - c)(c - a) \tag{3}$$

We are going to consider both approaches in the next sections corresponding to the growth of the lesion if no fibroblast is present. They are simple and generic models which have been extensively studied [14]. The first model, based on an unstable stationary value at $c = 0$, is related with the exponential growth of the bacteria

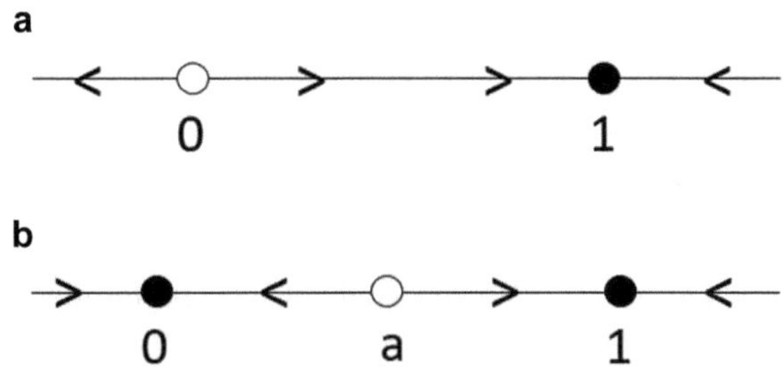

Fig. 3 Stability diagram for the unstable (**a**) and the stable (**b**) model. Stable points are represented in black and unstable points are represented in white for $0 < a < 1$. Arrows show the expected dynamics of the field c under intermediate initial values

inside the alveolus. The second model is related with a more realistic description of the propagation of the lesion among the alveoli, which is a complex dynamics.

2.2 Secondary Lobule Model

To simulate the secondary lobule, a three-dimensional (3D) model is created, which simulates the entire secondary lobule. The typical size of a secondary lobule is between 1 and 2.5 cm, while the mean size of an alveolus is 0.2 mm. Therefore, we have constructed a grid made up of $50 \times 50 \times 50$ alveoli, which produces a secondary lobule with $V = 1$ cm^3. For the reduction of the computational time we have also simulated a laminar section of the secondary lobule, using a two-dimensional (2D) system which consists of a 50×50 network of alveoli.

When the dependence on the space is included, fibroblasts present in the connective tissue appear surrounding the set of alveoli, simulating the pulmonary septa that packs the secondary lobule. Active motion of cells among the alveoli is added using a diffusion approach among neighbouring alveoli, see Fig. 2c. The diffusion of the bacilli to neighbouring regions that have not been infected is considered together with the spatial organizations of the alveoli, leading to the next dynamics for the fraction occupied by the lesion (c_l) in the tissue:

$$\frac{\partial c_l}{\partial t} = kc_l(1 - c_l)(c_l - a) + \nabla \cdot (D\nabla c_l), \tag{4}$$

where D is the diffusion coefficient of the active diffusion process of the macrophages, neutrophils, bacilli and other components of the lesion.

The fibroblasts are considered to only diffuse. The fraction occupied by the fibroblast and the collagen segregated during the encapsulation process (c_f) follows a diffusion equation:

$$\frac{\partial c_f}{\partial t} = \nabla \cdot (D_f \nabla c_f) \tag{5}$$

The diffusion coefficient of the fibroblast (D_f) is proportional to the modulus of the gradient and to the diffusion coefficient of the lesion:

$$D_f = \chi \cdot D \cdot |\nabla c_l|, \tag{6}$$

where χ is a constant. Therefore, we assume that there is a process of chemokinesis, where the diffusion increases with the gradient of cells forming the lesion. The fraction of volume occupied at the borders of the secondary lobule is imposed to be constant

The fraction occupied by the lesion (c_l) cannot occupy the space where fibroblast have segregated large amounts of collagen (c_f). This condition corresponds to a volume exclusion which decrease the diffusion under the presence of the other components. The total fraction of volume occupied by the lesion and the fibroblast together cannot be larger than one:

$$c_f + c_l \leq 1, \tag{7}$$

in the case of total fraction of volume equal to one, the whole space is completely full and no diffusion or dynamics are possible. Therefore we can define a new variable which takes into account the availability of space during the diffusion of both the lesion (c_l) and the fibroblast/collagen (c_f):

$$g = 1 - c_f - c_l. \tag{8}$$

This means that, for example, if an alveolus has been invaded by a fibroblast and the empty factor is 0, then the lesion will not be able to diffuse towards this direction. This variable also prevents fibroblasts to diffuse infinitely in an alveolus. The empty factor is always a number between $g = 1$, corresponding to empty space where cells can diffuse and enter and therefore the lesion grow and $g = 0$, corresponding full space and no diffusion or multiplication are possible. Given all these constraints, the model can be summarized into two equations:

$$\frac{\partial c_l}{\partial t} = kc_l(1 - c_l)(c_l - a) + \nabla \cdot (Dg\nabla c_l), \tag{9}$$

$$\frac{\partial c_f}{\partial t} = \nabla \cdot (\chi Dg|\nabla c_f| \cdot \nabla(c_f). \tag{10}$$

Furthermore, all the boundary points from the matrix corresponding to the fibroblasts are set to 1 at each step, keeping Dirichlet boundary conditions. To make sure there was not a problem with the empty factor, the borders of the matrix corresponding to the lesion have been set to 0.

The values of the parameters have been set to achieve a clear objective: to observe the process of encapsulation for each approach of the model. The comparison among

Table 1 Baseline values and units for 2D and 3D stable model and 3D unstable model

Parameter	Stable in 2D	Stable in 3D	Unstable in 3D	Units
a	0.16	0.12	–	–
D	$5 \cdot 10^{-4}$	$5 \cdot 10^{-2}$	$5 \cdot 10^{-5}$	mm^2/h
k	0.5	0.5	0.5	h^{-1}
χ	$1 \cdot 10^4$	$5 \cdot 10^2$	$1 \cdot 10^5$	–
ϵ	2.0	2.0	2.0	mm
α	5.0	7.0	3.0	–
Δt	$5 \cdot 10^{-4}$	$5 \cdot 10^{-4}$	$5 \cdot 10^{-4}$	h
Δx	0.2	0.2	0.2	mm
r_f	–	–	–	mm
τ_e	–	–	–	h

the values of the parameters in the single alveolus model and the corresponding 2D and 3D is shown in Table 1.

The 3D model was also tested in a non-rectangular lobule shape, see Fig. 4. To build it, we started from the six planes configuring a cube, and after that we defined eight planes with their director vectors and a point. Then, to construct the secondary lobule, we tested at each point the sign of the product between the director vector from the point to the plane, to detect at which side of each plane the point was. This allowed us to know if a point was inside the closed volume formed by all the planes, outside, or at the surface.

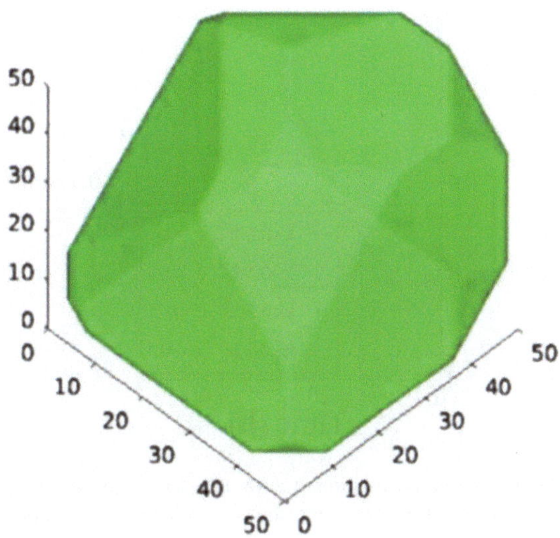

Fig. 4 Irregular shaped secondary lobule. This secondary lobule is made up of 97,878 alveoli, which gives a volume of $0.8\,cm^3$. The sphericity of this lobule is 0.58, with a major axis of 8.66 mm and a minor axis of 4.39 mm

3 Results

3.1 Single Alveolus Model

First we consider a single alveolus and, therefore integrate the equations without any spatial dependence. For the evolution of the dynamics of the fraction of cells, the Euler method has been employed for the numerical integration of the equations. For more details visit the repository: https://github.com/BIOCOM-SC/GranulomaFormation2024.git.

For the case of the unstable model, the fixed point at $c = 1$ is the only stable fix point of the system. The numerical simulations show that all trajectories end up in $c = 1$ for any value of the initial condition $c_0 > 0$, see Fig. 5a. The timescale of the process depends on the parameter k and it is determined by the factor k^{-1}.

There are two fixed points in the stable model, see Eq. (3). The temporal evolution of c is shown in Fig. 5b for different values of k and for different values of the initial condition c_0. The evolution shows that $c = 1$ and $c = 0$ are stable fixed points, while $c = a$ is an unstable fixed point. For this reason, trajectories whose initial condition is higher than the threshold $c_0 > a$ end up reaching the fixed point $c = 1$, whereas for initial conditions below the threshold $c_0 < a$ trajectories end up reaching $a = 0$. As in the previous case the timescale of the process is determined by the factor k^{-1}.

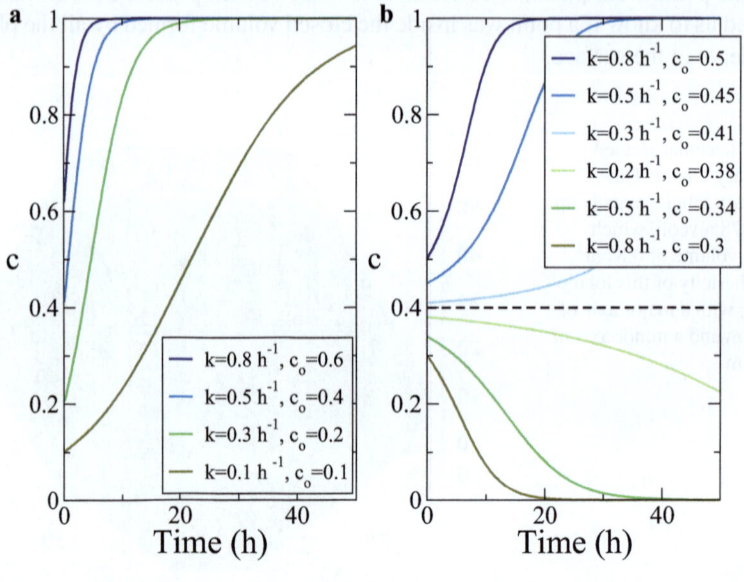

Fig. 5 Temporal evolution of c for the unstable model in Eq. (2) (**a**), and the stable model in Eq. (3), for $a = 0.4$ (**b**). Different values for k and initial conditions c_0 have been explored. Dashed line in panel (**b**) corresponds to the limit $c = 0.4$, stable solution with $a = 0.4$

3.2 Three-Dimensional Model of the Encapsulation

The three-dimensional model of the secondary lobule has been implemented in a $50 \times 50 \times 50$ grid (50×50 in the 2D case), where each point of the grid represents an alveolus of volume $8 \cdot 10^{-3}$ mm^3. For the evolution of the dynamics of the fraction of cells the Euler and spatial finite difference methods have been employed for the numerical integration of the equations. For more details visit the repository: https://github.com/BIOCOM-SC/GranulomaFormation2024.git.

3.2.1 Unstable Model

The numerical integration of the unstable model can show the process of encapsulation. However, in this case the evolution of the lesion is not stable and the interaction between the lesion and the fibroblasts causes the lesion to adopt irregular shapes. A non-spherical lesion is harder to encapsulate and therefore it has been difficult to achieve a case of encapsulation with this approach. The diffusion coefficient of the lesion had to be lowered and χ had to be increased by a few orders of magnitude. Also, the initial lesion was set very small in order to be able to see the encapsulation process. The Fig. 6 shows the eventual encapsulation of a lesion in a particular plane of axis z. There, the evolution of the lesion and fibroblast can be seen as well as the

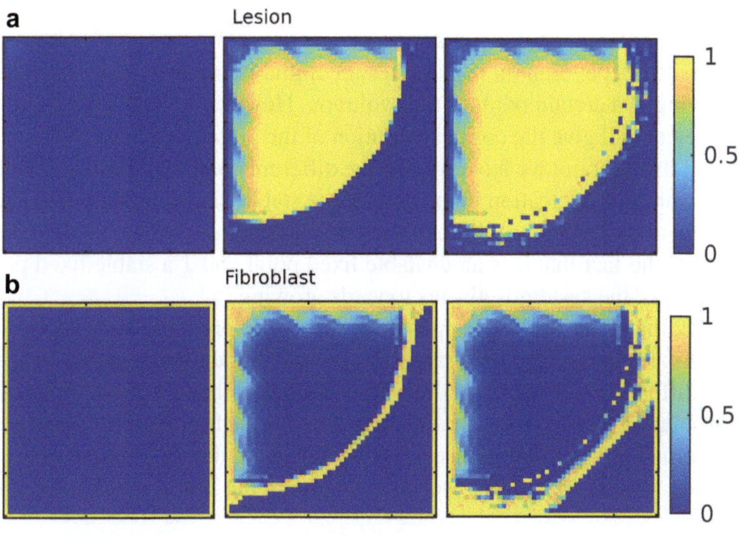

Fig. 6 Fraction of the volume occupied by the Lesion (**a**) and the fibroblasts and the collagen segregated (**b**) at three different moments of the evolution at z = 25, at three different moments of the evolution (time increases from left to right). Initial lesion was set to $\alpha = 3$ and $\epsilon = 2$ mm. The other parameters were set to: $D = 5 \cdot 10^{-2}$ mm^2/h, $\chi = 1.1 \cdot 10^5$ and $k = 0.2$ h^{-1}

empty factor, the diffusion coefficient of the fibroblast and the diffusion of both the lesion and the fibroblasts.

Such dynamics makes it difficult to use the model for the reproduction of the encapsulation process. The lesion may be encapsulated, however, the parameter values have to be fine-tuned and the resulting dynamics is, therefore, not robust to change of on the parameters. For this reason we employ also the stable model in the next section to model the encapsulation process.

3.2.2 Stable Model

The selection of the adequate initial condition is important because not every initial lesion configuration triggers the correct evolution of the lesion. The process of nucleation needs to take place in order to obtain the expected evolution of the lesion. Nucleation is the formation of an initial lesion that is able to continually grow. There are some initial conditions that cannot lead to the growth of the lesion and instead it will disappear. For this reason, the initial condition is defined as an hyperbolic tangent which depends on three parameters:

$$c = \frac{1}{2}\left[1 + \alpha \tanh\left(\frac{|\mathbf{r} - \mathbf{r}_0|}{\epsilon}\right)\right] \tag{11}$$

where α refers to the size of the initial lesion and ϵ refers to its sharpness. r_0 refers to the initial position of the lesion.

Not all the combinations of parameters ϵ and α lead to a correct evolution of the system. In fact, depending on their values the initial lesion eventually disappears or stops growing at a certain point of the evolution. However, to know which values of the parameters will give the correct evolution of the system is not as trivial as in the 0D case. In this section, we have studied the different combinations of parameters ϵ and α that can give nucleation in the case of the stable model, as shown in Fig. 7. In the unstable model, it is very difficult to obtain a case in which nucleation does not occur due to the fact that 0 is an unstable fixed point and 1 a stable fixed point, so the evolution of the system is always towards growing.

To check the effect of the parameters on the evolution of the system, we defined two output parameters: The final radius (r_f) of the granuloma, which is determined when the difference between two time steps is smaller than a certain threshold, and the encapsulation time (τ_e) corresponding to the time the fibroblasts begin to generate the granuloma until the final radius is determined. Note that the volume of the lesion is calculated at each step of the evolution assuming the lesion is spherical (circular in 2D).

After the nucleation of the lesion, it grows until arrives at the border where the encapsulation is achieved in the 3D stable model. In Fig. 8 we can see the evolution of the lesion in a plain of z axis. There, the evolution of the lesion and fibroblast can be seen as well as the empty factor, the diffusion coefficient of the fibroblast and the diffusion of both the lesion and the fibroblasts.

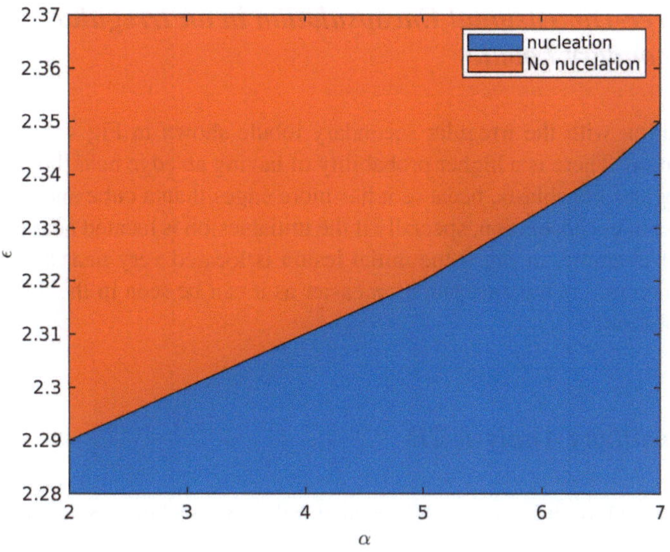

Fig. 7 Phase diagram for parameters ϵ and α. The red zone represents combinations that do not present nucleation, while blue zone presents nucleation

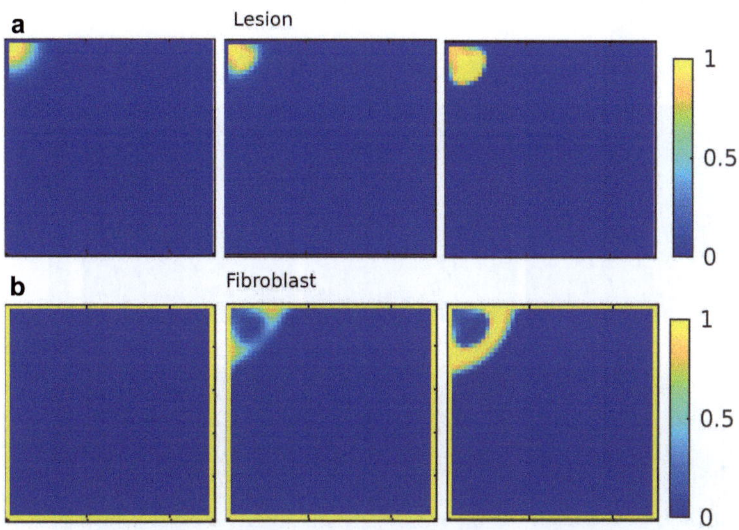

Fig. 8 Fraction of the volume occupied by the Lesion (**a**) and the fibroblasts and the collagen segregated (**b**) at three different moments of the evolution at z = 25, at three different moments of the evolution (time increases from left to right). Initial lesion was centered in [2, 2, 25], with $\alpha = 7$ and $\epsilon = 2.4$. The other parameters were set to:$D = 5 \cdot 10^{-2}$ mm^2/h, $\chi = 200$ and $k = 0.5$ h^{-1}

3.3 Three-Dimensional Encapsulation in an Irregular Secondary Lobule

The evolution with the irregular secondary lobule shown in Fig. 4 can also reach encapsulation. There is a higher probability of having an edge near the initial lesion in the irregular fibroblasts, because it has more edges than a cube and this can help the process of encapsulation, specially if the initial lesion is located near one of those edges. For example, in Fig. 9 the initial lesion is located very near an edge of the lobule which makes the encapsulation easier as it can be seen in the different steps of the evolution.

3.4 Sensitivity Analysis 2D

The 2D model has given successful results for the encapsulation process; the fibroblasts diffuse proportionally to the gradient of concentration of the lesion and stop the lesion from growing, forming a granuloma. Several steps of this process are shown in Fig. 10. At the beginning, a new infection is placed close to the connective tissue. When the lesion growths and affects the connective tissue, the fibroblasts begin to

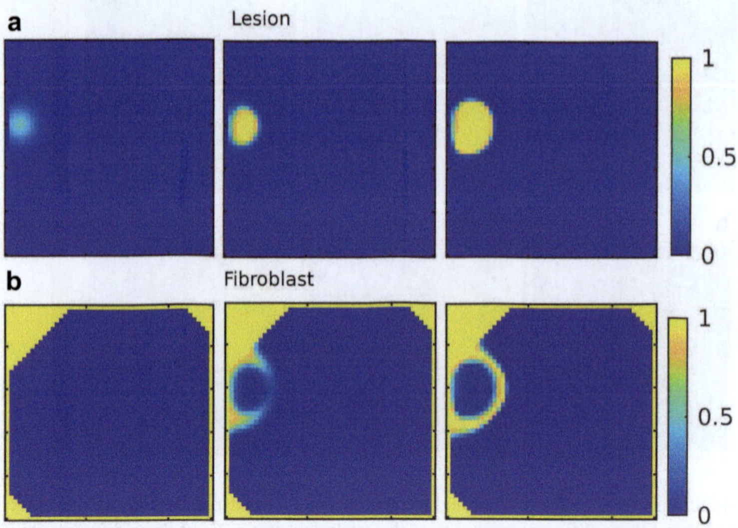

Fig. 9 Fraction of the volume occupied by the Lesion (**a**) and the fibroblasts and the collagen segregated (**b**) at three different moments of the evolution at z = 25, at three different moments of the evolution (time increases from left to right) and for z = 30. The shape of the secondary lobule is non-rectangular. Initial lesion was set to $\alpha = 5$ and $\epsilon = 2$. The other parameters were set to: $D = 5 \cdot 10^{-2}$ mm^2/h, $\chi = 200$ and $k = 0.5$ h^{-1}

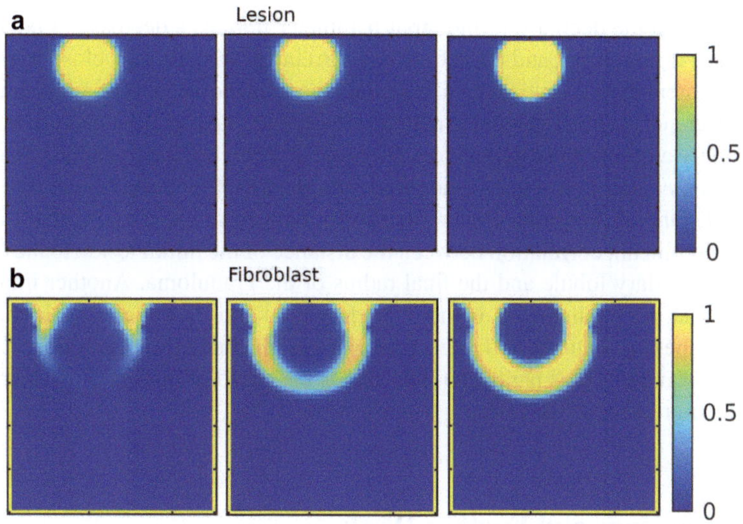

Fig. 10 Fraction of the volume occupied by the Lesion (**a**) and the fibroblasts and the collagen segregated (**b**) at three different moments of the evolution (time increases from left to right) in a regular two dimensional grid composed by 50 × 50 alveoli, giving rise to an area of 1 cm^2. Initial lesion was centered in [5, 20], with $\alpha = 5.0$ and $\epsilon = 2.0$. The other parameters were set to: $D = 0.5 \cdot 10^{-3}$ mm/h, $\chi = 100{,}000$ and $k = 0.5$ h^{-1}

diffuse from the borders of the secondary lobule due to the presence of the concentration gradient of the lesion. The granuloma is assumed to be formed when τ_e is reached.

A sensitivity analysis has been performed for some of the parameters of the model [9]. To do so, a methodology called Latin Hypercube Sampling has been used [13]: it consists of choosing a range for every parameter for which we want to study the influence on the results and divide each range into N points, N being, therefore, the number of values of each parameter we study. This way, the parameter space is divided into N^n boxes, where n is the number of parameters considered. These boxes are n-dimensional and they contain a small range of each of the parameters.

The algorithm to choose the N points is simple: a random box is chosen, and once chosen, the rows to which this box belongs cannot appear anymore. In fact, that is why we will end up with N points. Inside the box, a random point is chosen which determines the exact values of the n parameters.

In this case r_f and τ_e have been chosen to study their correlation with parameters k, d, D_f and the distance of the initial lesion to the closest fibroblasts wall. To do so, Partial Rank Correlation Coefficient (PRCC) has been used, which measures the linear dependence between one input parameter and one output parameter eliminating the influence of the rest of input parameters. In other words, if PRCC is positive the increase of the input parameter entails an increase of the output parameter, and in the other hand the decrease of the input parameter implies a decrease of the output parameter. If both input and output are not correlated, the PRCC is close to zero.

The computational time taken to calculate the whole evolution of a lesion until it is encapsulated in the 2D model is not very large, and that has allowed us to perform a sensitivity analysis from parameters D, χ and d. Their influence on the two output parameters, r_f and τ_e has been quantified with the PRCC and the results are shown in Fig. 11. One of the most remarkable results, as well as expected, is that Fig. 11a shows a significant correlation between the distance of the initial lesion to the closest wall of secondary lobule and the final radius of the granuloma. Another important result is that the parameter χ is the most decisive factor in the encapsulation time and the higher the distance to the secondary lobule, the more time the lesion will take to be encapsulated. This parameter determines the speed of motion of the fibroblast around the gradient of the lesion.

4 Conclusions and Further Work

A fast and effective encapsulation is important to determine if a patient will end up with an LTBI or if, instead, he will develop the active disease. We have built a mathematical model that reproduces the encapsulation of the granuloma. This model tries to bridge two different approaches: A complex and computationally demanding model based on partial differential equations for the details of the dynamics of the

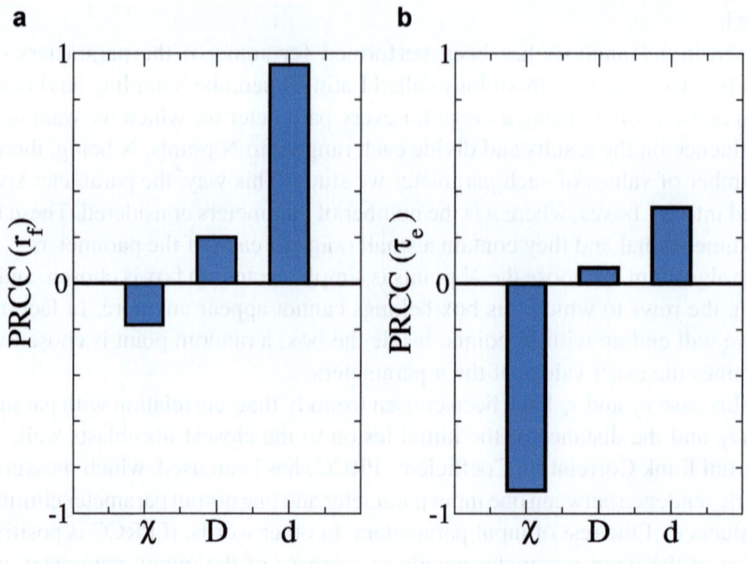

Fig. 11 Partial rank correlation coefficient from parameters D, χ and d for the quantities r_f (**a**) and τ_e (**b**). The number of samples of the Latin Hypercube Sample was $N = 1000$

Mtb inside the alveoli, and an Agent-based Model for the dynamics of the formed lesion on the spatial scale of the lung.

The 3D model reproduces correctly the evolution of a tuberculosis lesion, taking into account its exponential growth and the active diffusion of the cells among alveoli. The motion of fibroblasts from the pulmonary septa inwards and the chemokinesis behaviour are also successfully introduced in the model. Fibroblasts are able to stop the growth of the lesion, being able to trigger the formation of the granuloma. Relevant biological parameters are four: growth rate, diffusion coefficient of the lesion and the fibroblasts, and the threshold parameter a of the bistable model. In terms of computational time, there is a big improvement with respect to the reaction-diffusion model [6] which takes into account the whole interaction among different cells.

Very specific values of parameters and a very small initial lesion have to be set in order to finally see an encapsulation in the unstable approach. Obviously, this happens because of the lack of the second stable fixed point $c = 0$, making the stable approach the most useful and comfortable one to study the process of granuloma formation. Unstable dynamics appears naturally from the initial exponential growth of the number of bacilli. Is is probably necessary a threshold to initiate such growth from biological reasons although the mathematical solution is always an exponential growth. We include such threshold with the parameter a in the stable model.

However, it is mainly in the stable approach that encapsulation is produced. The results that the model has provided show that the distance of the initial lesion to the closest wall of secondary lobule is a relevant factor during the formation of the granuloma, which means that the initial infection location can be crucial to future development into latent tuberculosis infection or active tuberculosis. The distance to the wall is the critical parameter for the lesion, and therefore the probabilistic approach where the location of the lesion is randomly chosen has to be incorporated in the model to account for a more realistic description of the process.

The future work will consider the adaptation of this simple reaction-diffusion model to virtual models of the lungs. First a complete mapping between the parameters employed here and the more complex reaction-diffusion models to obtain adequate parameter values. The aim of building this model was to have a simple reaction-diffusion model to reproduce the results from the agent-based model. From the non-rectangular 3D lobule model, a whole 3D lung could be constructed by defining various non-rectangular lobules with different geometries and connecting them. The connection of the different lobules is done through a computational bronchial tree [5]. This would allow the study the dynamics of different initial lesions in the lungs, but with a reaction-diffusion evolution of the lesion and fibroblasts that potentially reproduce better the encapsulation dynamic.

Another interesting aspect of the model is the case in which the lesion is not encapsulated and therefore it invades the whole secondary lobule. The model we have built assumes that the connective tissue is a conserved envelope of the secondary lobule, however, this means that a lesion is eventually always encapsulated. Therefore, it would be interesting to reproduce the biological processes that make the lesion finally affect other secondary lobules in the non-encapsulation case.

Many other items could be added to the model to make it more realistic. For example, a non-homogeneous connective tissue could be implemented: having different fibroblasts densities would make the lobule structure much more realistic or an irregular distribution of alveoli inside the secondary lobule could be considered.

In summary, the simple reaction-diffusion model, introduced here, can be employed to extend previous results with more complex reaction-diffusion models applied to the volume of a secondary lobule into a virtual lung with all the secondary lobules simultaneously.

Acknowledgements We acknowledge the Spanish Ministry and Agencia Estatal de investigacion (AEI) for the Grant PID-2022-139216NB-I00 funded by MCIN/AEI/10.13039/501100011033 and by "ERDF A way of making Europe", by the "European Union".

References

1. Abar, S., Theodoropoulos, G.K., Lemarinier, P., O'Hare, G.M.: Agent based modelling and simulation tools: a review of the state-of-art software. Comput. Sci. Rev. **24**, 13–33 (2017)
2. Cardona, P.-J.: Revisiting the natural history of tuberculosis. Arch. Immunol. Ther. Exp **58**, 7–14 (2010)
3. Cardona, P.-J., Prats, C.: The small breathing amplitude at the upper lobes favors the attraction of polymorphonuclear neutrophils to Mycobacterium tuberculosis lesions and helps to understand the evolution toward active disease in an individual-based model. Front. Microbiol. **7**, 354 (2016)
4. Cardona, P.-J., Català, M., Arch, M., Arias, L., Alonso, S., Cardona, P., Lopez, D., Vilaplana, C., Prats, C.: Can systems immunology lead tuberculosis eradication? Curr. Opin. Syst. Biol. **12**, 53–60 (2018)
5. Català, M., Bechini, J., Tenesa, M., Pérez, R., Moya, M., Vilaplana, C., Valls, J., Alonso, S., López, D., Cardona, P.-J., Prats, C.: Modelling the dynamics of tuberculosis lesions in a virtual lung: role of the bronchial tree in endogenous reinfection. PLoS Comput. Biol. **16**(5), e1007772 (2020)
6. Català, M., López, D., Cardona, P.-J., Alonso, S., Prats, C.: A reaction-diffusion model to understand granulomas formation inside the secondary lobule during tuberculosis infection. PLoS One **15**(9), e0239289 (2020)
7. Hult, C., Mattila, J.T., Gideon, H.P., Linderman, J.J., Kirschner, D.E.: Neutrophil dynamics affect mycobacterium tuberculosis granuloma outcomes and dissemination. Front. Immunol. **12**, 712457 (2021)
8. Koch, A.J., Meinhardt, H.: Biological pattern formation: from basic mechanisms to complex structures. Rev. Mod. Phys. **66**(4), 1481 (1994)
9. Marino, S., Hogue, I.B., Ray, C.J., Kirschner, D.E.: A methodology for performing global uncertainty and sensitivity analysis in systems biology. J. Theor. Biol. **254**(1), 178–196 (2008)
10. Marzo, E., Vilaplana, C., Tapia, G., Diaz, J., Garcia, V., Cardona, P.J.: Damaging role of neutrophilic infiltration in a mouse model of progressive tuberculosis. Tuberculosis **94**(1), 55–64 (2014)
11. Pigozzo, A.B., Missiakas, D., Alonso, S., Dos Santos, R.W., Lobosco, M.: Development of a computational model of abscess formation. Front. Microbiol. **9**, 1355 (2018)
12. Prats, C., Vilaplana, C., Valls, J., Marzo, E., Cardona, P.J., López, D.: Local inflammation, dissemination and coalescence of lesions are key for the progression towards active tuberculosis: The bubble model. Front. Microbiol. **7**, 33 (2016)

13. Seaholm, S.K., Ackerman, E., Wu, S.C.: Latin hypercube sampling and the sensitivity analysis of a Monte Carlo epidemic model. Int. J. Bio-med. Comput. **23**(1–2), 97–112 (1988)
14. Scott, S.K., Showalter, K.: Simple and complex propagating reaction-diffusion fronts. Am. Chem. Soc. **96**(22) (1992)
15. Vilaplana, C., et al.: Granuloma encapsulation is a key factor for containing tuberculosis infection in minipigs. PLoS One **5**, e10030 (2010)
16. World Health Organization. Global tuberculosis report 2021. World Health Organization (2021)

Mathematical Model for Sustainable Biodegradation Process

Manuel Arrayás and Natalia González-Benítez

Abstract We present a technique to model the biodegradation processes of carbamazepine (CBZ) using a system of ordinary differential equations. The CBZ is a pharmaceutical compound with low solubility utilised worldwide and an emerging contaminant. Its biodegradation was mediated by the microbial consortium BOS08 extracted from wood rot in an Atlantic Forest. The mathematical model is obtained as a set of ordinary diferential equations, where a Picard contraction mapping technique is used to fix the value of the parameters involved by solving the inverse problem. The technique offers a powerful tool for simulating, predicting, and developing future models allowing biodegradation optimisation.

1 Introduction

In recent years, the emergence of newly identified contaminants, arising from both synthetic and natural sources, has introduced profound environmental challenges. These contaminants, characterized by their biological activity and notable resilience against inactivation or degradation processes, have given rise to a host of new environmental problems. The scientific exploration and modeling of the intricate interactions between these pollutants and microbial populations have evolved into a crucial and dynamic field of research. This article aims to provide an in-depth review of a specialized technique that we have employed to address the complex modeling of such systems, offering a powerful tool for simulating and predicting the outcomes of biological processes and facilitating the optimization of biodegradation strategies.

M. Arrayás (✉)
Área de Electromagnetismo, Universidad Rey Juan Carlos, Móstoles, Spain
e-mail: manuel.arrayas@urjc.es

N. González-Benítez
Departamento de Biología y Geología, Instituto de Investigación en Cambio Global (IICG-URJC), Física y Química Inorgánica, Universidad Rey Juan Carlos, Móstoles, Spain
e-mail: natalia.gonzalez@urjc.es

© The Author(s), under exclusive license to Springer Nature Switzerland AG 2025
M. Stich and J. Carballido-Landeira (eds.), *Nonlinear Dynamics for Biological Systems*,
SEMA SIMAI Springer Series 40,
https://doi.org/10.1007/978-3-031-99044-1_8

Given the multifaceted ways in which human activities adversely impact the environment, numerous recent studies have delved into the intricate linkages between anthropogenic actions and the surrounding ecosystems [1]. Presently, four prominent organizations—One Health, the Food and Agriculture Organization (FAO), the United Nations Environment Programme (UNEP), and the World Health Organization (WHO), in collaboration with the World Organization for Animal Health (WOAH)—operate within the One Health concept. This holistic approach integrates considerations of human, animal, agricultural, and ecosystem health to comprehensively address the challenges posed by the triple planetary crisis [2, 3]. This crisis, encompassing climate change, biodiversity loss, and environmental pollution, has spurred an intensified focus on research concerning emerging contaminants. The term "emerging contaminants" refers to a category of pollutants that includes pharmaceuticals and personal care products (PPCPs), ubiquitous in various environments worldwide. Even at trace concentrations (ng/L–μg/L), these PPCPs have been documented to cause adverse health impacts [4]. As a result, the scientific community is directing its attention towards understanding the behavior and effects of these contaminants, emphasizing the need for comprehensive research initiatives to address this critical aspect of environmental pollution. The current advances such as microalgae/fungal strains and anaerobic membrane bioreactors and membrane filtration combined with electrochemical oxidation processes are promising processes for the removal of these emerging contaminants [5] (Fig. 1).

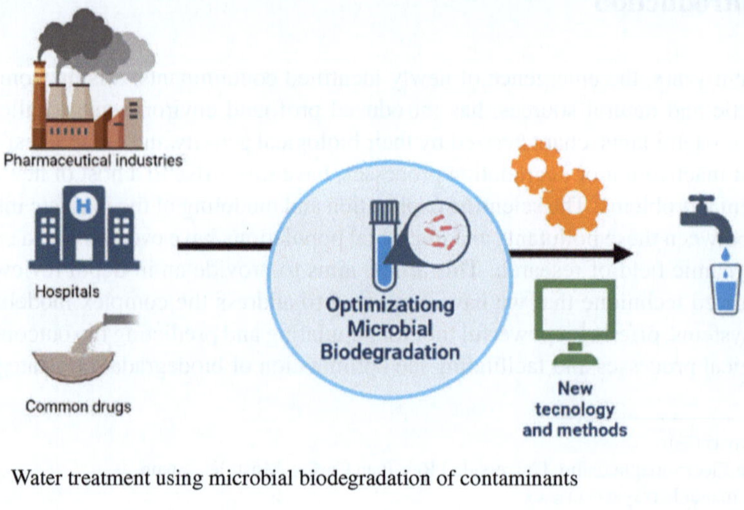

Fig. 1 Water treatment using microbial biodegradation of contaminants

1.1 The Biological Problem: CBZ Degradation by Microbial Consortium

Among numerous emergent pollutants, carbamazepine (CBZ) stands out as a highly effective antiepileptic drug, extensively detected in water systems due to its remarkable persistence. Numerous antecedent investigations have highlighted the suboptimal nature of most chemical, biological, and mechanical processes, resulting in a minimal biodegradation of CBZ [6]. It is crucial to note that comprehensive information regarding the diversity of microorganisms and enzymes capable of degrading CBZ remains limited [7]. Prevailing studies predominantly concentrate on individual strains [8], underscoring the existing gaps in our understanding.

In response to this lack of knowledge, the utilization of microbial consortia emerges as a promising strategy for enhancing the degradation of contaminants [9]. Unlike studies focused on individual bacterial isolates, microbial consortia offer several advantages in pursuing effective degradation processes. Employing the artificial method of pressure selection, microbial consortia can be meticulously optimized to degrade pollutants, effectively channeling the consortium towards the targeted pollutant [7]. A noteworthy feature of microbial consortia is their composition of multiple bacterial species, each equipped with distinct enzymes and metabolic pathways. This results in a broader metabolic diversity, thereby endowing the consortium with a more expansive capacity for degradation [10, 11].

The diverse species within microbial consortia imparts a heightened resilience to variations during biodegradation processes. These variations may manifest as changes in temperature, pH levels, nutrient and carbon availability, among other factors [12]. Such adaptability enhances the overall efficacy and robustness of microbial consortia in the face of dynamic environmental conditions.

This article specifically delves into the utilization of a microbial consortium denoted as BOS08, extracted from wood rot, and subsequently optimized using a selective pressure of CBZ to address the intricacies of the biodegradation process. The intentional selection and optimization of this consortium highlight the potential for advancing biodegradation studies through the strategic use of microbial consortia, offering a nuanced and resilient approach to combat the persistence of contaminants like CBZ in aquatic systems.

1.2 The Mathematical Problem

Since 1965, the study of pollutant biodegradation processes has gained significant attention, with over 40,000 articles published on the topic (Fig. 2, based on a Scopus search). The application of mathematical models to these processes has also been extensively explored, beginning in the 1970s and resulting in more than 1,000 publications to date. However, research on the biodegradation of emerging pollutants only began in the 2000s and has yielded fewer than 100 articles. Notably, only about

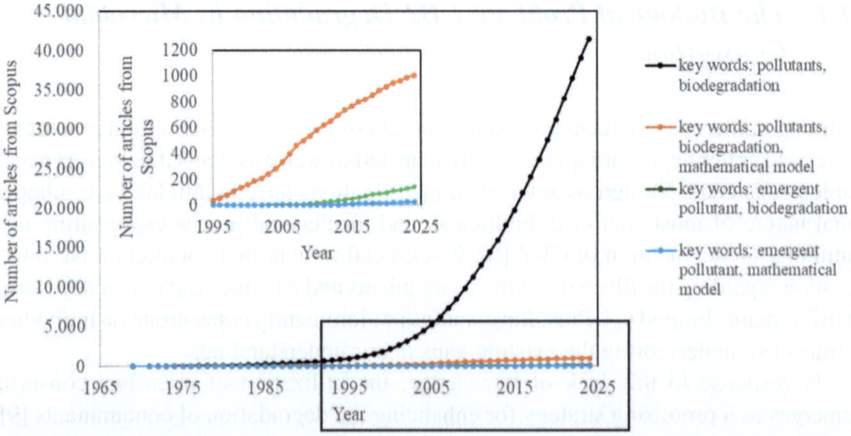

Fig. 2 Number of studies based on "biodegradation" and "pollutants" (black line). Number of studies based on "biodegradation", "pollutants" and mathematical model (orange line). Number of studies based on "biodegradation" and "emergent pollutants" (green line). Number of studies based on "emergent pollutants" and mathematical model (blue line). Number of studies based on "emergent pollutants" and ordinary mathematical equation (blue line). The exploring database used has been Scopus

twenty studies have employed mathematical models for optimizing the degradation of emerging pollutants (see Fig. 2), and none have utilized ordinary differential equation (ODE) models.

The ability to derive mathematical models of biodegradation processes brings about substantial advantages, facilitating significant time savings in experiments and expediting the quest for optimal concentrations. The transition from experimental data to a mathematical model depends on the specific model type under consideration. A recent review [13] shows how the computational approaches offer a cost-effective alternative for studying microbial communities through descriptive and engineering modelling. In this context, we posit the modelling of biodegradation dynamics as a deterministic system of ordinary differential equations (ODE). The primary objective is to ascertain an ODE system that faithfully replicates the empirical time series obtained from experimental data. This approach enables us to capture and understand the intricate dynamics of the biodegradation process quantitatively.

Assuming determinism in the ODE system allows us to explore and predict the biodegradation process's behaviour systematically. To formulate such a model, we rely on experimental measurements of bacterial population growth and drug concentration over time as crucial inputs. These measurements are the basis for fixing the model's parameters and aligning the model's predictions with observed data. The ensuing challenge involves solving an inverse problem for ordinary differential equations, where the objective is to deduce the appropriate set of equations that accurately reflect the empirical time series. It is important that incorporating any additional relevant knowledge into the modelling process can significantly enhance the reliability of the resultant model. Supplementary information, such as insights into the biochemical pathways involved, the impact of environmental factors, or the interplay between different microbial species, contributes to refining the model's accuracy. This holistic approach ensures a more comprehensive understanding of the biodegradation process and bolsters the predictive capabilities of the mathematical model. In summary, using mathematical models in biodegradation studies expedites experimental processes and allows for a systematic exploration of the dynamics involved. By conceptualizing the biodegradation process through a deterministic set of ODEs and integrating relevant experimental data, we can attain optimal concentrations and a more profound comprehension of the complex interplay between our consortium BOS08 and CBZ, specifically, as well as with other consortia, microorganisms, and substances in general.

1.3 The Mathematical Method to Obtaing the ODEs

To infer the ordinary differential equations (ODEs) that describe the dynamics of bacteria concentration $B(t)$ and contaminant concentration $C(t)$ over time, we need to consider the interactions between these two variables. Typically, the dynamics of such systems can be modeled by a set of coupled differential equations, where the rate of change of each variable depends on the other variable as well as possibly itself.

If we denote by B the bacteria concentration and by C the concentration of contaminants or pollutants, we can write in general

$$\frac{dB}{dt} = f_1(B, C), \quad B(0) = B_0, \tag{1}$$

$$\frac{dC}{dt} = f_2(B, C), \quad C(0) = C_0, \tag{2}$$

being f_1 and f_2 the generic functions to determine and B_0, C_0 the initial concentrations. The specific forms of $f_1(B, C)$, and $f_2(B, C)$ depend on the detailed mechanisms of interaction between bacteria and contaminants, which might include more complex terms if the system's biology or chemistry demands it.

We can further restrict the general form of the ODE system using the information provided by the underlying biological processes. Thus the ODE system will have a polynomial form, possibly only of first or second degree. Here we will take a second order polynomial approximation for the vector fields representing the evolution of our system. This approximation is done in the majority of the models employed for reaction kinetics and population-disease in biomathematical problems. Then, we propose the functional forms

$$f_1 = y_0 + y_1 C + y_2 B + y_3 C^2 + y_4 BC + y_5 B^2, \tag{3}$$

$$f_2 = x_0 + x_1 C + x_2 B + x_3 C^2 + x_4 BC + x_5 B^2. \tag{4}$$

As we will show, further biological insight will be able to restrict this polynomial form even more.

Once the biological information is incorporated into the model, the problem falls in the class of optimisation problems in which the task is to estimate the parameters appearing in the equations. In our case, the task is to find the kinetics coefficients x_i and y_i. Notice that in many practical situations, when the measurements are taken after a latent phase, the initial conditions B_0 and C_0 are also unknown and they have to be included as parameters to determine.

To get the parameters, we will take a least-square type algorithm for the optimisation problem, following closely the work by Kunze and Vrscay [14]. First we will integrate (1) and (2) to get

$$B(t) = \int_0^t f_1(B, C)\,dt + B_0, \tag{5}$$

$$C(t) = \int_0^t f_2(B, C)\,dt + C_0. \tag{6}$$

Then we take the square difference between our model solution and the experimental one and integrate over the interval $t \in [0, T]$, where from the experiments the concentrations $B_{exp}(t)$, $C_{exp}(t)$ are known,

$$\Delta_1^2 = \int_0^T |B_{exp}(t) - B(t)|^2\,dt,$$

$$\Delta_2^2 = \int_0^T |C_{exp}(t) - C(t)|^2\,dt. \tag{7}$$

The remaining task is to choose f_1, f_2 such that we get the minimum of Δ_1^2 and Δ_2^2. This corresponds to a minimum square optimisation problem. When the functions f_1, f_2 are restricted to a polynomial class, which is the case for (3) and (4), the optimisation problem has a solution [14].

1.4 The Algorithm for Two Dimensions

We will describe the recipe following [14]. The experimental data is normally available at discrete times. Thus we first need to interpolate the empirical data to get a solution $C_{exp}(t)$ and $B_{exp}(t)$ for any t in the interval $[0, T]$. Then we substitute (3) and (4) into (5) and (6)

$$B(t) = \int_0^t \left(y_0 + y_1 C_{exp} + y_2 B_{exp} + y_3 C_{exp}^2 + y_4 B_{exp} C_{exp} + y_5 B_{exp}^2\right) dt + B_0, \tag{8}$$

$$C(t) = \int_0^t \left(x_0 + x_1 C_{exp} + x_2 B_{exp} + x_3 C_{exp}^2 + x_4 B_{exp} C_{exp} + x_5 B_{exp}^2\right) dt + C_0. \tag{9}$$

We can write those expressions as

$$B(t) = \sum_{n=0}^{5} y_n g_n(t) + B_0, \tag{10}$$

$$C(t) = \sum_{n=0}^{5} x_n g_n(t) + C_0, \tag{11}$$

where the definitions of $g_n(t)$ follow from (8) or (9).

Now we can build the error functions

$$\Delta_1^2 = \int_0^T |B_{exp}(t) - B_0 - \sum_{n=0}^{5} y_n g_n(t)|^2 \, dt, \tag{12}$$

$$\Delta_2^2 = \int_0^T |C_{exp}(t) - C_0 - \sum_{n=0}^{5} x_n g_n(t)|^2 \, dt, \tag{13}$$

and make the corresponding functional variation $\delta\Delta_1^2/\delta B_0$, $\delta\Delta_1^2/\delta y_i$, $\delta\Delta_2^2/\delta C_0$, $\delta\Delta_2^2/\delta x_i$, with $i = 1 \ldots n$ to get the stationary conditions when they are equal to zero. Notice that we treat the initial values B_0 and C_0 as variables to determine from the variational problem. In many biological problems, the initial concentration is unknown as there is a latency time before the reactions start. Thus we get a linear system to solve for each set which reads

$$\mathbf{G}\mathbf{p}_{1,2} = \mathbf{a}_{1,2}, \tag{14}$$

being $\mathbf{G}$ a symmetric matrix,

$$\mathbf{G} = \begin{bmatrix} 1 & \langle g_0 \rangle & \cdots & \langle g_5 \rangle \\ \langle g_0 \rangle & \langle g_0 g_0 \rangle & \cdots & \langle g_0 g_5 \rangle \\ \vdots & \vdots & \ddots & \vdots \\ \langle g_5 \rangle & \langle g_5 g_0 \rangle & \cdots & \langle g_5 g_5 \rangle \end{bmatrix}, \tag{15}$$

with

$$\mathbf{p}_1 = \begin{bmatrix} B_0 \\ y_0 \\ \cdots \\ y_5 \end{bmatrix}, \quad \mathbf{p}_2 = \begin{bmatrix} C_0 \\ x_0 \\ \cdots \\ x_5 \end{bmatrix}, \tag{16}$$

and

$$\mathbf{a}_1 = \begin{bmatrix} \langle B_{exp} \rangle \\ \langle g_0 B_{exp} \rangle \\ \cdots \\ \langle g_5 B_{exp} \rangle \end{bmatrix}, \quad \mathbf{a}_2 = \begin{bmatrix} \langle C_{exp} \rangle \\ \langle g_0 C_{exp} \rangle \\ \cdots \\ \langle g_5 C_{exp} \rangle \end{bmatrix}. \tag{17}$$

The bracket notation means here $\langle h \rangle = \int_0^T h(t)\,dt$. We observe that the matrix G is the same for both sets of parameters in the variational problem, so it only needs to be calculated once. The optimal value of the parameters are found by solving (14) for $\mathbf{p}_1$ and for $\mathbf{p}_2$.

2 Application of the Method: The Degradation Process of CBZ by the Microbial Consortium BOS08

In recent days, the predictive biology represents one step more to the synthetic biology in microbiology. To designing complex multi-species bacterial communities with specific role are becoming plausible thanks to the application of the mathematical models [15].

We will present in this section a case illustrating the mathematical method described in the previous one. We will include only the polynomial terms which can be justified from biological considerations and take a role in the degradation of CBZ by the microbial consortium BOS08 [7].

2.1 The Model

In this example we present how we modelled the action of the microbial consortium BOS08 [7] on the CBZ. The data were obtained from laboratory experiments to record the evolution of both the CBZ concentration and the microbial consortium. We will denote by C and B the concentrations of CBZ and bacteria consortium respectively and write assume that the dynamics of the system is given by (1) and (2).

We could use the general forms (3) and (4), but let us restrict them. First, for the bacteria growth rate we will take a logistic term of the form $B(\beta_2 - B)$. We recall that a logistic term takes into account the fact that the bacterial growth slows down when microbial density gets so large that resources become limited.

To model the interaction between CBZ and the bacteria consortium, we need to consider the experimental fact that the bacteria concentration has to reach a critical concentration before they can start the degradation of CBZ. So we will include a term proportional to $C(B - \beta_1)$. We will further make use of the Monod hypothesis that tells that any homogeneous interacting term appears in both equations but opposite in sign [16]. Them in the CBZ equation rate there will be a term proportional to $-C(B-\alpha_1)$. Note that by mass conservation, the parameters α_1 and β_1 get close values if our hypothesis that the degradation of the CBZ is mainly due to the direct action of the bacterias is right.

Moreover, further experiments revealed that CBZ did not present any abiotic degradation or endogent metabolism, so no further terms will be included and we write up the following model for our system,

$$\begin{aligned} f_1(B,C) &= \beta_0[C(B-\beta_1)+B(\beta_2-B)], \\ f_2(B,C) &= -\alpha_0 C(B-\alpha_1). \end{aligned} \qquad (18)$$

The terms considered are polynomial terms present in the second order approximations (3) and (4). We want to point that including further terms would lead to a better fitting of the experimental series, but with the risk of overfitting the model. We also stress that all terms included are biologically justified.

2.2 *Fitting of the Model to Experimental Data*

We can now proceed to implement of the algorithm outlined in Sect. 1.4 to get the coefficients of the model (18). In the Figs. 3 and 4 we can see the experimental data obtained from measurements in the laboratory at different time intervals. The concentration of the bacterias is measured by optical absorption so it is given in arbitrary units.

We use the empirical data to feed the model and calculate the matrix G and the right hand side of (14). Then we can solve for the parameters of the model and write then in the form given by (18). The parameter values are written in the Table 1.

In agreement with our initial guess, the fitted values of α_1 and β_1 are very close, confirming the hypothesis that the bacteria concentration needs to reach a critical concentration before the CBZ degradation phase starts to growth exponentially. Also,

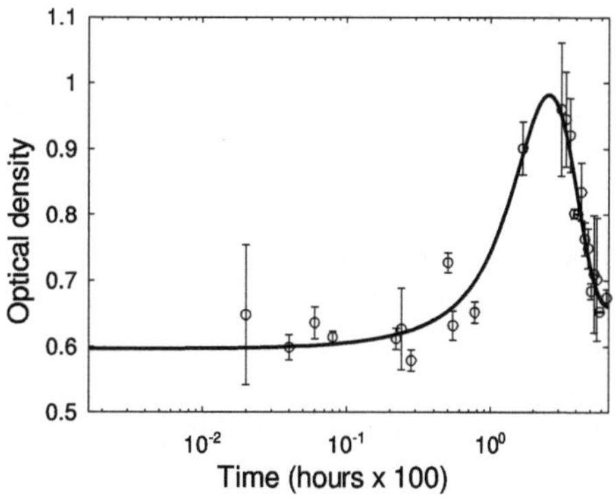

Fig. 3 Bacterias concentration versus time. The bacterial concentration is measured as optical density in arbitrary units. Notice the log scale in time. Points are empirical data. The continuous curve is obtained integrating the model

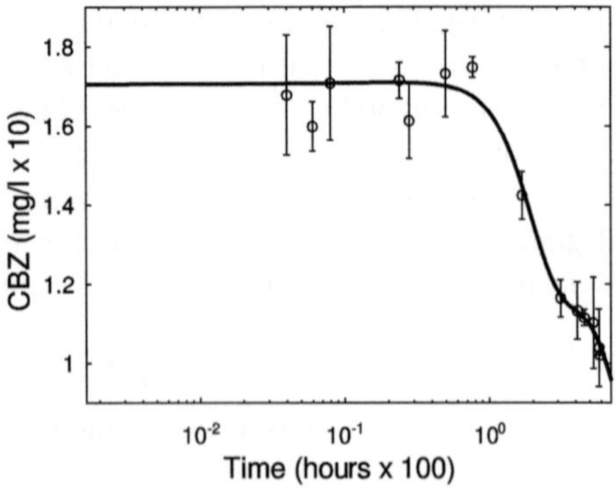

Fig. 4 Carbamazepine concentration versus time in log scale. Points are empirical data and the continuous curve is obtained by numerical integration of the model

Table 1 The parameters model obtained by the fitting algorithm

f_1	f_2
$B_0 = 0.5955$	$C_0 = 1.6427$
$\beta_0 = 2.3807$	$\alpha_0 = 0.3092$
$\beta_1 = 0.4953$	$\alpha_1 = 0.5387$
$\beta_2 = 0.3426$	

the initial concentrations are calculated from the data as stated before. This is very convenient as the first measurements are taken after an initial time interval in the experiments.

In order to test our model, we can solve and plot it together with the experimental data. In the Figs. 3 and 4 the continuous lines are obtained by the numerical integration of our model with the parameters and initial conditions given in the Table 1. We found a good agreement between the theoretical model and the experimental data. Further discussion and interpretation of values found for the parameters of the model can be found in [7].

3 Conclusions

3.1 Towards Synthetic Microbiology

Due to the complexity of the biodegradation process, constructing mathematical models combined with experiments using microbial consortia may provide a framework to understand, predict and maximise biodegradation processes [17]. These models can help describe the kinetics of pollutants biodegradation as a function of factors such as the evolution of the nutrient and carbon concentration, microbial growth rates, and the variability of the abiotic parameters. Through model simulations and sensitivity analyses, it is possible to obtain the key factors driving the degradation processes, identify rate-limiting steps, and propose strategies to enhance the process efficiency. The predictive capability of the mathematical models is valuable in assessing the feasibility and effectiveness of biodegradation strategies from laboratory-scale experiments before implementation to real-world applications.

In particular, differential equation models play an importal role for the building of computational models with predictive capabilities. Here we have presented a method to obtain the parameters when the model is given by a set of ODEs of polynomial type. We have formulated the fitting of the parameters as an inverse problem and use Picard contraction mapping to get a solution. Then we have illustrated the method for a case study of the biodegradation process of CBZ by a microbial consortium.

3.2 Applications in the Industry

Mechanistic models, particularly kinetics-based models utilizing ODE systems, are well-suited for elucidating observed phenomena, predicting process behavior, and analyzing inherent bioprocess characteristics, including controllability [18]. Consequently, industries can use these models for bioprocess monitoring, facilitating the reduction of experimental efforts and resource expenses while ensuring the sustained maintenance of desired conditions, reproducibility, and suitability for application in biotechnological production [19]. A recent area of research employing mathematical models is synthetic biology, which helps in bioengineer cells to create new biological pathways and novel components.

Recent advances in genomic research and genetic engineering have enabled the creation of genetic devices and biological modules [20]. These innovations allow industries to design and optimize microorganisms for the biodegradation of pollutants or biosynthesis of biofuels, pharmaceuticals, and other valuable compounds, with ecological friendliness [19]. The current challenge in synthetic biology lies in simplifying complex biological data into mathematical models that can simulate and predict outcomes by linking synthetic DNA sequences to dynamic phenotypes [21].

In this context, synthetic biology relies on omics approaches such as genomics, transcriptomics, proteomics, and metabolomics, which are well-suited for discovering new species, enzymes, and metabolic pathways capable of acting on a wide range of compounds. These discoveries hold significant potential for applications in the chemical, agrochemical, pharmaceutical, food, and biochemical industries. To handle this data and to deal with the complexity of the results, bioinformatics tools must combine biological principles with mathematical models, statistics and machine learning [22].

Most laboratory experiments must be scaled up to large-scale industrial processes using mathematical models [23], which serve as versatile tools for optimizing biological processes, enhancing efficiency, reducing costs, and promoting sustainability. This approach has significantly improved manufacturing across a wide range of products in the chemical, agrochemical, pharmaceutical, food, and biochemical industries. In summary, mathematical models offer a powerful and flexible toolset for understanding, optimizing, and controlling biotechnological and industrial processes, resulting in greater efficiency, cost reduction, and sustainability [24].

Acknowledgements This work was funded by Universidad Rey Juan Carlos, Programa Propio ETHERNET (Estructura y dinámica de redes con interacciones de orden superior) M3038.

References

1. Häder, D.P., et al.: Anthropogenic pollution of aquatic ecosystems: emerging problems with global implications. Sci. Tot. Environ. **713**, 136586 (2020)
2. Atlas, R.M.: One Health: its origins and future. Curr. Top. Microbiol. Immunol. **365**, 1–13 (2013)
3. Mackenzie, J.S., Jeggo, M., Daszak, P., Richt, J.A.: One Health: the human-animal-environment interfaces in emerging infectious diseases. Curr. Top. Microbiol. Immunol. **366** (2013)
4. Kumar, M., et al.: Current research trends on emerging contaminants pharmaceutical and personal care products (PPCPs): a comprehensive review. Sci. Tot. Environ. **859**, 160031 (2023)
5. Shahid, M.K., Kashif, A., Fuwad, A., Choi, Y.: Current advances in treatment technologies for removal of emerging contaminants from water—a critical review. Coord. Chem. Rev. **442**, 213993 (2021)
6. Chen, X., et al.: Removal processes of carbamazepine in constructed wetlands treating secondary effluent: a review. Water **10**(10), 1351 (2018)
7. González-Benítez, N., Molina, M.C., Arrayás, M.: Empirical evidence and mathematical modelling of carbamazepine degradative kinetics by a wood-rotting microbial consortium. Waste Biomass Valor. **12**, 995 (2021)
8. Moghaddam, A., et al.: Biodegradation of pharmaceutical compounds in industrial wastewater using biological treatment: a comprehensive overview. Int. J. Environ. Sci. Technol. **20**(5), 5659–5696 (2023)
9. Ahmed, P.M., de Figueroa, L.I., Pajot, H.F.: Application of microbial consortia in degradation and detoxification of industrial pollutants. In: Microorganisms for Sustainable Environment and Health, pp. 401–418 (2020)
10. Brenner, K., You, L., Arnold, F.H.: Engineering microbial consortia: a new frontier in synthetic biology. Trends Biotechnol. **26**(9), 483–489 (2008)
11. Roell, G.W., et al.: Engineering microbial consortia by division of labor. Microb. Cell Factor. **18**, 1–11 (2019)
12. Zhou, Q., et al.: Interactions among microorganisms functionally active for electron transfer and pollutant degradation in natural environments. Eco-Environ. Health **2**, 3–15 (2023)
13. García-Jiménez, B., Torres-Bacete, J., Nogales, J.: Metabolic modelling approaches for describing and engineering microbial communities. Comput. Struct. Biotechnol. J. **19**, 226–246 (2021)
14. Kunze, H., Vrscay, E.: Solving inverse problems for ordinary differential equations using the Picard contraction mapping. Inverse Probl. **15**, 745–770 (1999)
15. Lopatkin, A.J., Collins, J.J.: Predictive biology: modelling, understanding and harnessing microbial complexity. Nat. Rev. Microbiol. **18**(9), 507–520 (2020)
16. Monod, J.: The growth of bacterial cultures. Annu. Rev. Microbiol. **3**, 371–394 (1947)
17. Wade, J.M.: Not just numbers: mathematical modelling and its contribution to anaerobic digestion processes. Processes **8**(8), 888 (2020)
18. St. Amand, M.M., Tran, K., Radhakrishnan, D., Robinson, A.S., Ogunnaike, B.A.: Controllability analysis of protein glycosylation in CHO cells. PLoS One **9**(2), e87973 (2014)
19. Luo, Y., Kurian, V., Ogunnaike, B.A.: Bioprocess systems analysis, modeling, estimation, and control. Curr. Opin. Chem. Eng. **33**, 100705 (2021)

20. Benner, S.A., Sismour, A.M.: Synthetic biology. Nat. Rev. Genet. **6**(7), 533–543 (2005)
21. Kaznessis, Y.N.: Models for synthetic biology. BMC Syst. Biol. **1**, 1–4 (2007)
22. Djordjevic, M., Rodic, A., Graovac, S.: From biophysics to omics and systems biology. Eur. Biophys. J. **48**, 413–424 (2019)
23. Janoska, A., Buijs, J., van Gulik, W.M.: Predicting the influence of combined oxygen and glucose gradients based on scale-down and modelling approaches for the scale-up of penicillin fermentations. Process Biochem. **124**, 100–112 (2023)
24. Molina, M.C., et al.: Bioremediation techniques for naproxen and carbamazepine elimination. Chimica Oggi-Chem. Today **34**(2), 52–55 (2016)

MIX
Papier aus verantwortungsvollen Quellen
Paper from responsible sources
FSC® C105338

If you have any concerns about our products,
you can contact us on
ProductSafety@springernature.com

In case Publisher is established outside the EU,
the EU authorized representative is:
Springer Nature Customer Service Center GmbH
Europaplatz 3, 69115 Heidelberg, Germany

Printed by Libri Plureos GmbH
in Hamburg, Germany